Advanced Computational Approaches for Water Treatment

A rapid growth in global industrialization and population has triggered intense environmental pollution that has led to a water crisis, resulting in the decay in the quality of human life and economic losses. Novel water purification techniques are expected to alleviate this challenge. Recently, various water purification techniques, along with different computational techniques, have been developed. For instance, water purification techniques such as electromagnetic water purification, solute–surface interactions in water, micro-magnetofluidic devices, UV-led water purification, and use of membranes can be thoroughly investigated by using a range of computation techniques, such as molecular dynamics, the lattice Boltzmann method, and the Navier–Stokes method-based solver. *Advanced Computational Approaches for Water Treatment: Applications in Food and Chemical Engineering* presents these different numerical techniques and traditional modeling and simulation approaches to elaborate on and explain the various water purification techniques.

Features:

- Serves as a dedicated reference for this emerging topic
- Discusses state-of-the-art developments in advanced computational techniques for water purification
- Brings together diverse experience in this field in one reference text
- Provides a roadmap for future developments in the area

This book is primarily intended for chemical engineers, hydrologists, water resource managers, civil engineers, environmental engineers, food scientists and food engineers interested in understanding the numerical approaches for different water purification techniques, such as membrane, sedimentation, filtration, micromagnetofluidic device, and ozone/UV, among others.

Advanced Computational Approaches for Water Treatment

Applications in Food and Chemical Engineering

Edited by
Krunal M. Gangawane, Madhuresh Dwivedi,
and Praveen Ghodke

CRC Press
Taylor & Francis Group
Boca Raton London New York

CRC Press is an imprint of the
Taylor & Francis Group, an **informa** business

First edition published 2024
by CRC Press
6000 Broken Sound Parkway NW, Suite 300, Boca Raton, FL 33487-2742

and by CRC Press
4 Park Square, Milton Park, Abingdon, Oxon, OX14 4RN

CRC Press is an imprint of Taylor & Francis Group, LLC

Library of Congress Cataloging-in-Publication Data
Names: Gangawane, Krunal, editor. | Dwivedi, Madhuresh, editor. | Ghodke, Praveen, editor.
Title: Advanced computational approaches for water treatment : applications in food and
chemical engineering / edited by Krunal M. Gangawane, Madhuresh Dwivedi, Praveen Ghodke.
Description: Boca Raton : CRC Press, 2024. | Includes bibliographical references and index.
Identifiers: LCCN 2023013052 (print) | LCCN 2023013053 (ebook) |
ISBN 9781032350141 (hardback) | ISBN 9781032350691 (paperback) |
ISBN 9781003325147 (ebook)
Subjects: LCSH: Water–Purification–Technological innovations. |
Water–Purification–Simulation methods. | Computational fluid dynamics
Classification: LCC TD477 .A37 2024 (print) |
LCC TD477 (ebook) | DDC 620.1/0640285–dc23/eng/20230627
LC record available at https://lccn.loc.gov/2023013052
LC ebook record available at https://lccn.loc.gov/2023013053

ISBN: 9781032350141 (hbk)
ISBN: 9781032350691 (pbk)
ISBN: 9781003325147 (ebk)

DOI: 10.1201/9781003325147

Typeset in Times
by Newgen Publishing UK

Contents

Preface

Water management in a convenient environment remains an important problem for several influencing variables. An accurate and deterministic forecasting of water treatment with respect to concentration of pollutants, water temperature, pollutant dosage, and methodology provides a wide variety of schematizations, erroneous measurements, and prior assumptions to treatments. In consideration of the exponential growth of recent industry and wastewater results in an methodology for an optimal treatment system of such a water effluent system to mitigate the environmental impact with respect to regulatory standards. This leads to the need for innovatation in wastewater control management to treat residual waters in search of environment. Accurate and reliable runoff prediction models should be a core alternative for effective water effluent management. In addition to water resource management, the challenges and uncertainties due to climate changes and artificial interference, computational approaches provide a sustainable environment in modeling water resource variables for the decision making process.

Computational approaches have ascended as a powerful solution for reconnoitering water treatment phenomena and solving real-life problems. Currently, there are several computational approaches allowed in complex phase analysis; continuous and discrete have been the two important aspects in any of the computational approaches so far. In the continuous approach, the governing equations can be obtained by the application of conservation of mass, momentum, and energy over an infinitesimal control volume. These equations are further discretized by using suitable discretization technique. Consequently, the recovered set of algebraic equations are then solved by some applied numerical method. On the other hand, the discrete approach concentrates on mimicking the molecular movement within the system (which are expressed by equations like Boltzmann's equation and Hamilton's equation). However, both approaches have pros and cons, and continuous development/improvement in the existing computational methods has been taking place.

In recent years, the world has witnessed a rapid growth in the field of computational approaches owing to its abundant benefit in the water effluent industry. The relevance of advanced computational methods has helped in understanding the fundamental physics of thermal and hydrodynamics behavior of fluids that can provide benefits to the water treatment industry in enormous applications. Therefore, in this book the use of different numerical/computational techniques for the simulation of fluid flow, heat and mass transfer in several water treatment processes is outlined.

The intent of the book is to provide a single information source for readers interested in the use of methods based on the numerical/computational analysis as applied in water treatment and technology. It covers the whole spectrum and applications from water resource management to industrial water effluent systems. The book also explores various numerical procedures used for modeling and validation in water treatment systems. It can be a ready reckoner for the academician/researcher to learn the advanced numerical technique in water treatment operations.

Editor Biographies

Krunal M. Gangawane is Assistant Professor of Chemical Engineering at Indian Institute of Technology Jodhpur (IIT Jodhpur), India. He has done his graduation in Chemical Engineering (B.Tech.) from the University of Pune in 2007. Later, he received his M.Tech. and Ph.D. degrees in Chemical Engineering from Indian Institute of Technology Roorkee in 2010 and 2015, respectively. His Ph.D. research was based upon the development of CFD code based on the lattice Boltzmann method for convective heat transfer problems. His current area of research is magneto-convection, nanofluidics, enhanced oil recovery, aerogels, etc. He has more than 20 publications in journals of repute. He joined UPES Dehradun as Assistant Professor in 2015 and, subsequently, moved to NIT Rourkela in March 2018 as Assistant Professor in the Chemical Engineering Department. Recently, (March 2023) He joined IIT Jodhpur. Dr. Gangawane received 'Best Researcher-Individual Excellence–December 2017' during MANTHAN-2017 at UPES Dehradun. He was chosen as a Member of Academia-Industry interaction (2016) in UPES for conducting research at Reliance R&D. He had a research collaboration with the Firat University (Turkey), Ghent University (Belgium), King Saud University (Saudi Arabia) on the topic of 'Convection heat transfer in enclosed bodies for different fluids.' He is also the recipient of the ISRO-RESPOND sponsored project in March 2020. Dr. Gangawane is the reviewer of various peer-reviewed international journals.

Madhuresh Dwivedi, M.Tech., Ph.D., is Assistant Professor in Department of Food Process Engg of National Institute of Technology (NIT) Rourkela. Dr. Dwivedi completed B. Tech. degree in Agricultural Engineering from College of Agricultural Engineering, Jabalpur India, in the year 2010, M. Tech. in Food Process Engineering from Indian Institute of Technology (IIT), Kharagpur in the year 2012 and Ph.D. (2015) at Department of Food Process Engineering, Indian Institute of Technology (IIT) Kharagpur.

Praveen Ghodke has a Ph.D. from the Indian Institute of Technology Bombay, 7 years of expertise in research and teaching, and a strong experimental background in the use of waste to energy (WtE). He has led a projects with a 2-DST grant and a patent on a catalyst for the pyrolysis of mixed plastic waste. Outstanding mentorship was recognised with the 'Dronacharya Award,' and corresponding students were given the 'Eklavaya Award.' His work was presented at various national and international symposiums, and he also authored more than 40 articles that were peer-reviewed and published as articles, conference proceedings, and book chapters. He was chosen to be an associate member of IChemE-UK and a life member of IIChE.

Contributors

Richa Agarwal
Department of Chemical Engineering
Banasthali Vidyapith, Rajasthan,
    India

Chandra Shekar Bestha
Department of Chemical Engineering
National Institute of Technology
    Calicut, Kozhikode, Kerala,
    India

Madhuresh Dwivedi
Department of Food Process
    Engineering
National Institute of Technology
    Rourkela, Odisha, India

Krunal M Gangawane
Department of Chemical Engineering
National Institute of Technology
    Rourkela, Odisha, India

Praveen Kumar Ghodke
Department of Chemical Engineering
National Institute of Technology
    Calicut, Kozhikode, Kerala, India

Ravikant R. Gupta
Department of Chemical Engineering
Banasthali Vidyapith, Rajasthan,
    India

Naval Koralkar
Department of Chemical Engineering
    at ITM (SLS) Baroda University
    (ITMBU), India

Abhishek Kumar
Department of Chemical Engineering
National Institute of Technology
    Rourkela, Odisha, India

Brajesh Kumar
Department of Chemical Engineering
National Institute of Technology
    Srinagar, India

Pradeep Kumar
Department of Chemical Engineering
Institute of Engineering & Technology
    Lucknow, U.P., India

Sudhanshu Kumar
Department of Chemical Engineering
National Institute of Technology
    Rourkela, Odisha, India

Vineet Kumar
Department of Chemical Engineering
Indian Institute of Technology-ISM
    Dhanbad, Jharkhand, India

Mahendran Radhakrishnan
Centre of Excellence in Non-Thermal
    Processing,
National Institute of Food Technology,
    Entrepreneurship and Management
    (NIFTEM)–Thanjavur
Tamil Nadu, India

Anbarasan Rajan
Centre of Excellence in Non-Thermal
    Processing
National Institute of Food Technology,
    Entrepreneurship and Management
    (NIFTEM)–Thanjavur
Tamil Nadu, India

E.J. Rifna
Department of Food Process
    Engineering
National Institute of Technology
    Rourkela, Odisha, India

Chingakham Ngotomba Singh
Department of Food Process
  Engineering
National Institute of Technology
  Rourkela, Odisha, India

Vijay Singh
Department of Chemical Engineering
Institute of Engineering & Technology
  Lucknow, U.P., India

P Suhailam
Department of Chemical Engineering
National Institute of Technology
  Calicut, Kozhikode, Kerala,
  India

Niranjan Thota
Department of Food Process
  Engineering
National Institute of Technology
  Rourkela, Odisha, India

G.S.N.V.K.S.N. Swamy Undi
Air Ok Technologies Private Limited,
  Research Park,
Indian Institute of Technology, Chennai,
  Tamil Nadu, India

Raju Yerolla
Department of Chemical Engineering
National Institute of Technology
  Calicut, Kozhikode, Kerala, India

1 Advanced Computational Approaches for Water Treatment

Abhishek Kumar[1] and Krunal M. Gangawane[1,2]*
[1]Department of Chemical Engineering, National Institute of Technology Rourkela,
Rourkela, Odisha, India
[2]Department of Chemical Engineering, Indian Institute of Technology Jodhpur,
Jodhpur, Rajasthan, India
*Corresponding Author

CONTENTS

1.1 INTRODUCTION

Computational fluid dynamics (CFD) as a tool for conducting process analysis of fluid flows is becoming more used in various sectors. It has recently gained widespread recognition for analyzing hydraulic issues in water and wastewater treatment (WWT); according to Samstag et al. (2016), significant cost and risk factors support using CFD for improved wastewater design. However, only a few academic organizations specialize in teaching or researching CFD in the wastewater industry, and only a limited number of recommendations are available. As a result, a committee of workers within the International Water Association (IWA) Specialist Group Modelling and Integrated Assessment (MIA) was established to encourage the formation of a group like this one (*International Water Association–International Water Association*, n.d.) (Laurent et al., 2014). This group provided an alternative explanation method of utilizing CFD as a dietary complement to utilizing more straightforward models and focused on effective modeling techniques for CFD modeling of WWT (Alex et al., 2002)(Takht Ravanchi et al., 2009).

The field of computational fluid dynamics, often known as CFD, uses complex numerical models to make predictions about flow, mixing, and (bio)-chemical reactions. In the field of engineering that deals with drinking water, CFD is being used increasingly to optimize treatment systems and to provide predictions about how well such installations will work. CFD modeling is a powerful tool for understanding drinking water systems' hydrodynamic and (bio)-chemical processes. Supposing it has been adequately applied, considering the turbulent regions' complicated movements has been supported by experiments (Wols, 2011).

Understanding the hydrodynamic fluid characteristics of a porous adsorption filter fluid contained in a vertical tube may be facilitated using CFD using ANSYS as an instrument. The characteristics of pressure and flow rate are included in this category of hydrodynamic fluids. It was discovered that the bench-scale filter had a state that was neither uniform nor radially asymmetric when the pressure was applied. The gradient contour plot by Kumar (Kumar et al., 2022) helped us understand the bench-scale filter's limitation concerning the flow phase, resulting in a low flow rate within the filter. The parameters determined from the bench-scale computational study, which included the permeability constant and the loss coefficient, helped determine the scale-up dimension at which the flow regime displayed an uneven gradient contour. The significance and relevance of computational fluid dynamics in establishing the scaled-up dimensions of a filter column. The CFD method can potentially be advantageous for setting the groundwork for scale-up investigations (Lim, 2021).

The term CFD refers to a method that uses numbers to compute the characteristics of a fluid in motion. Most procedures for treating water include moving the water around in some way. Keeping up with this motion may be very challenging and time-consuming, not to mention costly. If flow patterns and other aspects of fluids in motion could be predicted, it would give insight into processes that would not have been conceivable without such knowledge (Norton & Sun, 2006; Samstag et al., 2016).

On the other hand, there is not yet an examination of the individual processes, including a historical review of work focused on creating specific unit process methods. The purpose of this research is to provide a comprehensive evaluation of the current state of the art regarding the use of CFD to assess water and WWT facilities, as well as to supply a vital evaluation of the research requirements that will be required in the future.

1.1.1 MOLECULAR DYNAMICS FOR MEMBRANES PROCESS

It is because there are more and more worries about water availability. In the last several decades, membrane-based water treatment methods have become required to overcome these challenges since they can enhance how well and fast water is treated. It is because membranes can filter out impurities from water. Also, the processes that utilize membranes are simple to implement and may be amplified to meet the needs of the industry. They are also very energy efficient. There are several membrane technologies used for water purification nowadays. They can be classified according to their primary motivating factor: the pressure difference (microfiltration (MF), ultrafiltration (UF), nanofiltration (NF), reverse osmosis (RO)), concentration difference (forward osmosis (FO)), electric potential difference (electro dialysis (ED), electrode ionization (EDI), and temperature difference (membrane distillation). Due to the growing use of membrane technology in water treatment processes, it has become imperative to increase our knowledge of synthetic membranes and separation mechanisms (Ebro et al., 2013).

Utilizing the computer method known as molecular dynamics (MD) is one approach to achieving this goal. Classical MD simulations offer benefits over other computer simulation approaches because they can compute parameters linked to a system's dynamic, such as transport coefficients, time-dependent fluctuations, reactions to system changes, rheological characteristics, and spectra (Schlick, 2010).

MD calculates these values by merging mathematics, chemistry, physics, and computer science principles. MD employs computer simulations to solve the Newtonian equations of motion numerically. It explores physical processes and movement in systems at the molecular level, often at the nanometer size, by computing equilibrium and dynamic parameters that are usually difficult to evaluate using straightforward analytical or experimental approaches. Due to these factors, MD is gaining prominence in membrane research, whose operations are based on the molecular motions of its components (She et al., 2008). In the coming chapter, we have to study further the molecular dynamics of the membrane process.

1.1.2 MEMBRANE PROCESS FOR WATER PURIFICATION: CFD APPROACH

Membrane technology is increasingly employed in water treatment. Membrane technology is reliable, low-energy, high-efficiency, simple to run, and maintains no phase change, unlike evaporation. Membrane filtration-increasing single-element and module-based systems are developed and manufactured for various businesses. For market competitiveness, compact and flexible membrane-filtering systems improve. However, flexible and compact designs might hinder filtering. Optimizing between compactness, flexibility, and system performance depends on membrane design. Module layout in a filtering system incorporating pipelines generates further pressure loss, which a superior design may prevent (Othman et al., 2021).

CFD tools can model a membrane filtration system's pressure, velocity, and temperature. Various processes need different simulations. Some can be addressed using regular means, while others need specific procedures (Wiley & Fletcher, 2002). CFD allows quick, straightforward, adaptable, and affordable membrane filtering system design optimization (Moulin, 2015).

CFD tools are user-friendly, flexible, and cost-effective. These features assist in analyzing and optimizing membrane systems. CFD may assess simulation factors, such as module shape, filtration pipe, pipe connection, membrane permeability, and module number. Appendix A introduces the calibration-library-interpolation (CLI) scale, and porous media approximation methodologies used in this study (Oliveira Neto et al., 2021). It streamlined the simulation model and made the CFD tool PC-friendly. Due to similar membrane system architecture, CLI might be used. Various libraries were produced for the modules, connection, pipe, and connector geometries. CFD simulation generates pressure loss and velocity data for all substructures. In this method, the whole membrane filtering system must not be designed in 3D, saving time. With CLI, a system simulation may be done with known parameters, like volume flow or pressure using libraries and interpolation. Geometry setup and parameter changes are more straightforward than conventional CFD simulation (Ghidossi et al., 2006).

1.1.3 MICROMAGNETOFLUIDIC DEVICE

This section explains a potential method for removing heavy metals from water using magnetic nanoparticles in microfluidic water flow devices as a component of the purification process (Kefou et al., 2016). In this method, the micromixer is the essential equipment, while external magnetic fields make the mixing and driving particles so effective (Cai et al., 2017).

Nanoparticles, microfluidics, and magnetohydrodynamics are coupled for water purification. Due to the low Reynolds number (no turbulence), microscale mixing is sluggish. A magnetohydrodynamic (MHD) micromixer speeds chemical reaction rates instead of using massive water tanks to agitate nanoparticles and heavy metals to tackle this challenge (Mitra & Chakraborty, 2012). Micromixers boost micromixing and are popular. They are active or passive. Passive mixers do not need external actuation to increase mixing; they employ microchannel designs, have no moving components, and use no energy. Homogeneous mixing is caused by molecular diffusion or chaotic advection. External disturbances help active mixers mix. Electrohydrodynamic, pressure-driven, dielectrophoretic, electrokinetic, magnetohydrodynamic, acoustic, thermal, etc. Active micromixers generate heterogeneous mixing by convection (Ward & Fan, 2015). This discovery offers a microfluidic system to get magnetic nanoparticles (MNPs) and heavy metals near enough to react quicker in water microchannels chemically.

Hydrodynamic interactions between particles and heavy metals create motion. This system is a micromagnetofluidic structure that produces attracting and repulsive forces. In a microchannel, particles are slowed by the walls and form horizontal eddies. External magnetic forces help MHD micromixers achieve good flow micromixing. The applied current, the electromagnet's geometry, the magnetic particles' magnetization moments, their size, and the microchannel's location relative to the electromagnet all affect micromixing. Figure 1.1 depicts the MHD micromixer being researched. A macro-sized external permanent electromagnet acted on magnetic particles. By switching the electromagnets on and off, the micromixer could produce periodic magnetic forces that agitated the magnetic particles to oscillate along the microchannel. Fluids and magnetic particles enhanced mixing efficiency (Kamali et al., 2014; Kefou et al., 2016).

A computational approach for modeling magnetism particles coated with suitable metal oxides to extract heavy metals from an aqueous solution in microchannels; the method would be used in the simulation of magnetic particles. A magnetically powered micromixer arrangement is thought to improve mixing even when there is

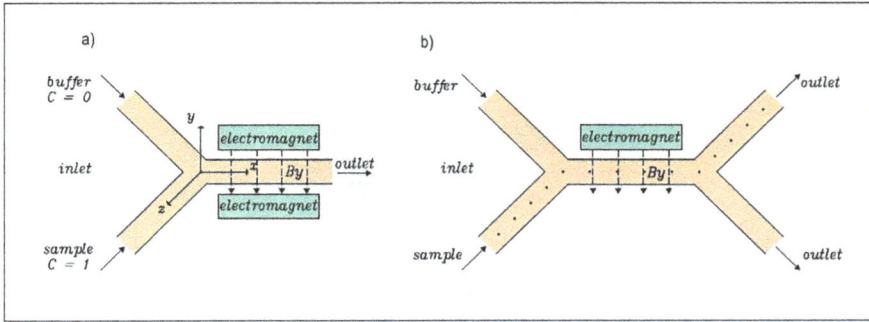

FIGURE 1.1 Water purification in micromagnetofluidic devices: mixing in MHD micromixers (Kefou et al., 2016).

very little turbulence in the system. In the current computational research, we first solve flow field calculations using the 2D and 3D Navier–Stokes equations to replicate the mixing process. A Lagrangian approach is used to model the motions of the magnetic nanoparticles (Pamme, 2006).

To enhance the velocity and trajectory of particles when subjected to various magnetic fields from the outside and to improve the blending efficiency in a fluid environment. The magnetic field had the optimal gradient, which made it easier for particles to navigate into the required track and stay away from the channel walls. Meanwhile, the flow field is shown. Compared to the conventional approaches, it is anticipated that this technology will increase chemical speed and decrease the time required for water purification (Wang et al., 2008).

1.1.4 COAGULANT TREATMENT

In the process of coagulation, iron or aluminum salts like aluminum sulfate, ferric sulfate, ferric chloride, or polymers are added to the water. These chemicals, which have a positive charge, are called coagulants. The positive charge of the coagulant cancels out the negatively charged particles in the water. When this happens, the particles stick together. It is called coagulation. The larger particles, called floc, are heavy and quickly sink to the water's bottom. The process by which things settle is called sedimentation. Figure 1.2 shows the coagulation's basic steps (Safe Drinking Water Foundation, 2021).

A facility that treats water will first add the coagulant to the water before swiftly mixing the two. It will ensure that the coagulant is distributed evenly throughout the water. The water that has been coagulated may either be filtered immediately via a medium filter (such as sand and gravel), a settling tank, or a microfiltration or ultrafiltration membrane. Each of these options has its advantages and disadvantages. The larger particles sink to the bottom of a settling tank, also known as a clarifier, where they are collected for further removal. The water is then ready to proceed to the filtering stage of the treatment process (Hahn et al., 2012).

WATER COAGULATION

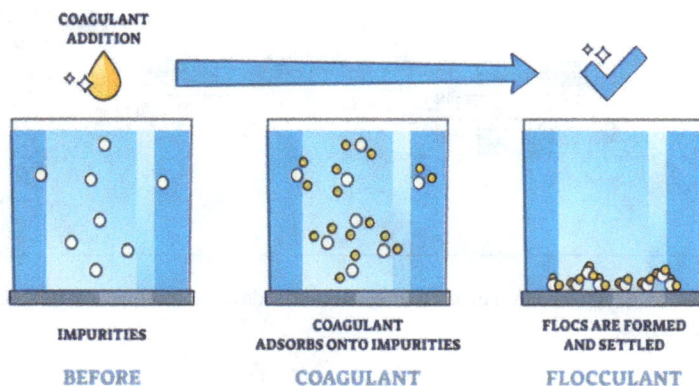

COAGULANT
ADDITION

| IMPURITIES | COAGULANT ADSORBS ONTO IMPURITIES | FLOCS ARE FORMED AND SETTLED |
| BEFORE | COAGULANT | FLOCCULANT |

FIGURE 1.2 Different types of coagulants may be used for the treatment of water (Greenwood, 2020).

CFD is used to examine mixing in coagulation and flocculation processes. Due to the limitless speed of competing for chemical reactions, WWT coagulation is particularly sensitive to mixing. Using CFD, one may examine the impact of various quick mix and flocculation designs and ascertain the needed level of mixing. Flow modeling and visualization may prevent the velocity gradients likely to result in floc breaking (Birjandi et al., 2013).

The next step in the modeling process would be to add flocs and fluid interactions into the CFD models. Several options exist, ranging from complete Eulerian–Eulerian two-phase modeling to one-phase modeling, provided that the impact of the floc phase on the liquid phase is minimal. To develop a broader CFD flocculation model, it is also necessary to include the chemical interactions between various species.

1.1.5 Electromagnetic Method

Electromagnetic fields, sometimes known as EMFs, have been shown to have potentially beneficial effects on water and wastewater treatment. The primary benefits of EMFs over other types of fields are their safety, compatibility, and simplicity, as well as their cheap operating costs, environmental friendliness, and lack of known adverse consequences. These technologies have been used traditionally to remediate wastewater. Not just the antibacterial agent but also EMFs have been shown to have antimicrobial effects on wastewater, and these fields possess other beneficial qualities changing the physical and chemical characteristics of the purpose of the wastewater treatment process molecules of water and other elements, the precipitation of sludge, phosphorus, and organic compounds. Some of these properties include the elimination of chemicals found in wastewater. This research will examine the recent developments that have been made in the use of EMFs in the treatment of sewage and wastewater, as well as the underlying working mechanics. Furthermore, prospects of technology related to this field include the field being analyzed and considered (Yadollahpour et al., 2014).

1.1.6 PHOTOCATALYTIC REACTOR

Conventional wastewater treatment techniques are ineffective in removing dangerous organic compounds' pollutants; thus, an expanding variety of novel treatment technologies are now being studied and developed. One of them, the study of heterogeneous photocatalysis, is a rapidly developing and promising technology (Herrmann, 2005). This method has the potential to mineralize harmful molecules (carbon dioxide, water, and mineral acids) into innocuous compounds (carbon dioxide, water, and mineral acids) without creating additional waste streams (Ollis et al., 1991).

Regardless of the numerous advantages of photocatalysis and the significant experimental studies on this subject, several challenges must be overcome before large-scale photocatalytic oxidation reactors can be developed for water treatment. The lack of appropriate models and simulation tools for forecasting and assessing the performance of full-scale systems, and therefore the absence of sufficient scale-up methods, is one of the primary challenges impeding the commercialization of water treatment systems based on this technology (Adesina, 2004; Mukherjee & Ray, 1999). Adopting computational fluid dynamics (CFD), which has been thoroughly proven to be a highly successful tool in designing, developing, and scaling-up reactive systems, is an excellent method for addressing this problem (Ranade, 2002). Through the simultaneous modeling of hydrodynamics, species mass transport, chemical reaction kinetics, and photon flux distribution, CFD enables a full investigation of the performance of photocatalytic reactors by giving the local values of the parameters of interest (i.e., fluid velocity, pollutant concentration, reaction rate, UV irradiance, etc.). Furthermore, applying CFD analysis to scaled-up reactors reduces experimental effort and manufacturing costs at the pilot-scale level (Boyjoo et al., 2014).

Few CFD experiments have been conducted on photocatalytic water treatment reactors. Two studies have been done on turbulent flow immobilized photocatalytic reactors for water treatment depending on the researchers' understanding (Sengupta et al., 2001) are mentioned in the literature. In the first, several simulations were carried out to research the influence that flow rates, diffusion coefficients, reaction rate constants, and inter-lamp separation had on the functioning of a multiamp reactor. According to the research, the reactor's conversions were mostly governed by the contaminant's flow and diffusion, not the surface reaction rate (Ray, 1998). However, neither the articles nor the simulation findings are compared to experimental data.

Duran et al. (2011) describe the findings of the other study. This work investigated several hydrodynamic (laminar and Reynolds stress models [RSMs]) to anticipate mass transfer from the outside and oxidation caused by photocatalysis. The assessment used a flat-plate differential reactor and formic acid as a model chemical. Based on simulations of the system's radiation field, the UV irradiance above the reactor's photocatalyst plates was considered constant. As a result, irradiance modeling was omitted from the model and not assessed. When entirely laminar or turbulent flow regimes were present in the system, there was a high correlation between the model predictions and the actual results (Rodriguez et al., 2009).

The primary objective of this research was to get an understanding, via experimentation, of a CFD-based model that can be used to simulate photocatalysts reactors that have water treatment surfaces covered with nanoparticles. The computer model combines hydrodynamics, species mass transfer, chemical reaction kinetics, and irradiance distribution within the reactor. These factors have been investigated separately, and several intake layouts, dimensions, and light sizes will be evaluated experimentally in the forthcoming chapters.

1.2 ADVANTAGES AND DISADVANTAGES OF COMPUTATIONAL TECHNIQUES FOR WATER PURIFICATIONS

Advantages:

1. CFD uses a numerical technique to solve the difficult momentum, heat, and mass equations concurrently; as a result shown in Figure 1.3, it may open new windows to a better understanding of what is occurring throughout the water treatment process.
2. Identifying and optimizing temperature hot spots and concentration polarizations is one of the many benefits of using CFD modeling rather than more traditional approaches. Because of this, the impacts of temperature and polarization may be determined and reduced with CFD models, particularly when it comes to modules with complicated shapes.
3. The biggest problem is the unavailability of wastewater treatment modules and membranes. Understanding the MD process and simulating transport phenomena may assist. Thus, designing, building, and testing modules takes time and money. Modeling may save time, energy, and money.

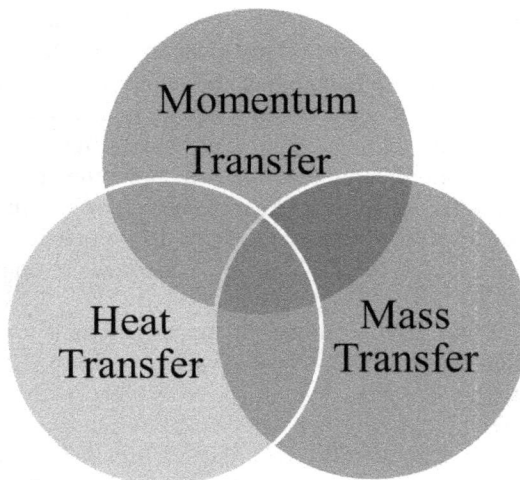

FIGURE 1.3 CFD uses a numerical approach.

4. It does not require much water for re-circulation in the condensing unit.
5. The standard distillery water treatment procedure is inefficient, expensive, and requires maintenance. It also requires long detention, which lengthens therapy. Soft computing may solve this constraint. This enables the designer to simulate different conditions.
   - Many flows and heat transfer processes cannot be easily tested.
   - CFD allows excellent control over the physical process and provides the ability to isolate specific phenomena for study.
6. A CFD model can simulate the hydrodynamics of a design before implementation.
   - Reducing lead-up time and costs.
   - Ultimately leading to the optimization of reactor configuration.

Disadvantages:

1. Highly sensitive to the various operational settings that were employed.
2. Sometimes not consistently supported by the results of experiments: the primary concern of CFD applications to industrial plants is choosing the optimum geometry and operating conditions to obtain the best hydrodynamic and mass transfer performances.
3. A large amount of processing power is needed to run some computational problems.
4. The input data may entail some guesswork or imprecision, so CFD simulations may never be certain of their results. The mathematical model does not have a mechanism that is adequate to represent the issue.
5. Physical models
   - The computational fluid dynamics method is based on physical representations of actual world processes.
   - The accuracy of CFD solutions is limited to that of the underlying physical models.
6. Numerical errors (due to approximation in the numerical models).
7. Boundary conditions
   - Comparable to actual physical models.
   - The initial boundary conditions given to the numerical models can only be as accurate as the CFD solution used to solve the problem.
8. CFD can now achieve this goal thanks to computer power. Enhancing the simulation requires further effort. Understanding hydrodynamics' physical phenomena are connected. It requires supplementary experimental investigation.

REFERENCES

Adesina, A. A. (2004). Industrial exploitation of photocatalysis: progress, perspectives, and prospects. *Catalysis Surveys from Asia, 8*(4), 265–273.
Alex, J., Kolisch, G., & Krause, K. (2002). Model structure identification for wastewater treatment simulation based on computational fluid dynamics. *Water Science and Technology, 45*(4–5), 325–334. https://doi.org/10.2166/wst.2002.0616

Birjandi, N., Younesi, H., Bahramifar, N., Ghafari, S., Zinatizadeh, A. A., & Sethupathi, S. (2013). Optimization of coagulation-flocculation treatment on paper-recycling wastewater: application of response surface methodology. *Journal of Environmental Science and Health, Part A, 48*(12), 1573–1582.

Boyjoo, Y., Ang, H., & Pareek, V. (2014). CFD simulation of a pilot scale slurry photocatalytic reactor and design of multiple-lamp reactors. *Chemical Engineering Science, 111*, 266–277. https://doi.org/10.1016/j.ces.2014.02.022

Cai, G., Xue, L., Zhang, H., & Lin, J. (2017). A review on micromixers. *Micromachines, 8*(9). https://doi.org/10.3390/mi8090274

Duran, J. E., Taghipour, F., & Mohseni, M. (2011). Evaluation of model parameters for simulating TiO_2 coated UV reactors. *Water Science and Technology, 63*(7), 1366–1372.

Ebro, H., Kim, Y. M., & Kim, J. H. (2013). Molecular dynamics simulations in membrane-based water treatment processes: A systematic overview. *Journal of Membrane Science, 438*, 112–125. https://doi.org/10.1016/j.memsci.2013.03.027

Ghidossi, R., Veyret, D., & Moulin, P. (2006). Computational fluid dynamics applied to membranes: State of the art and opportunities. *Chemical Engineering and Processing: Process Intensification, 45*(6), 437–454. https://doi.org/10.1016/j.cep.2005.11.002

Greenwood, J. (2020). *How are coagulants and flocculants used in water and wastewater treatment.* WCS Group. www.wcs-group.co.uk/wcs-blog/coagulants-flocculants-wastewater-treatment

Hahn, H. H., Hoffmann, E., & Odegaard, H. (2012). *Chemical Water and Wastewater Treatment VI.* Springer Science & Business Media.

Herrmann, J.-M. (2005). Heterogeneous photocatalysis: state of the art and present applications. *Topics in Catalysis, 34*(1–4), 49–65. https://doi.org/10.1007/s11244-005-3788-2

Kamali, R., Shekoohi, S. A., & Binesh, A. (2014). Effects of magnetic particles entrance arrangements on mixing efficiency of a magnetic bead micromixer. *Nano-Micro Letters, 6*(1), 30–37. https://doi.org/10.1007/BF03353766

Kefou, N., Karvelas, E., Karamanos, K., Karakasidis, T., & Sarris, I. E. (2016). Water purification in micromagnetofluidic devices: Mixing in MHD micromixers. *Procedia Engineering, 162*, 593–600. https://doi.org/10.1016/j.proeng.2016.11.105

Kumar, P., Brar, S. K., & Cledon, M. (2022). A computational fluid dynamics approach to predict the scale-up dimension of a water filter column. *Case Studies in Chemical and Environmental Engineering, 5*(March), 100201. https://doi.org/10.1016/j.cscee.2022.100201

Laurent, J., Samstag, R. W., Ducoste, J. M., Griborio, A., Nopens, I., Batstone, D. J., Wicks, J. D., Saunders, S., & Potier, O. (2014). A protocol for the use of computational fluid dynamics as a supportive tool for wastewater treatment plant modelling. *Water Science and Technology, 70*(10), 1575–1584. https://doi.org/10.2166/wst.2014.425

Lim, Y. (2021). Computational Fluid Dynamics (CFD) of Chemical Processes. In *ChemEngineering.* MDPI. https://doi.org/10.3390/books978-3-03943-934-8

Mitra, S. K., & Chakraborty, S. (2012). *Microfluidics and Nanofluidics Handbook: Fabrication, Implementation, and Applications.* Taylor & Francis.

Moulin, P. (2015). *Computational Fluid Dynamics (CFD) and Membranes BT–Encyclopedia of Membranes* (E. Drioli & L. Giorno (Eds.); pp. 1–2). Springer Berlin Heidelberg. https://doi.org/10.1007/978-3-642-40872-4_142-1

Mukherjee, P. S., & Ray, A. K. (1999). Major challenges in the design of a large-scale photocatalytic reactor for water treatment. *Chemical Engineering & Technology: Industrial Chemistry-Plant Equipment-Process Engineering-Biotechnology, 22*(3), 253–260.

Norton, T., & Sun, D.-W. (2006). Computational fluid dynamics (CFD)–an effective and efficient design and analysis tool for the food industry: A review. *Trends in Food Science & Technology*, *17*(11), 600–620. https://doi.org/10.1016/j.tifs.2006.05.004

Oliveira Neto, G. L., Oliveira, N. G. N., Delgado, J. M. P. Q., Nascimento, L. P. C., Magalhães, H. L. F., Oliveira, P. L. de, Gomez, R. S., Farias Neto, S. R., & Lima, A. G. B. (2021). Hydrodynamic and performance evaluation of a porous ceramic membrane module used on the water–oil separation process: an investigation by CFD. *Membranes*, *11*(2), 121. https://doi.org/10.3390/membranes11020121

Ollis, D. F., Pelizzetti, E., & Serpone, N. (1991). Photocatalyzed destruction of water contaminants. *Environmental Science & Technology*, *25*(9), 1522–1529.

Othman, N. H., Alias, N. H., Fuzil, N. S., Marpani, F., Shahruddin, M. Z., Chew, C. M., David Ng, K. M., Lau, W. J., & Ismail, A. F. (2021). A review on the use of membrane technology systems in developing countries. *Membranes*, *12*(1), 30. https://doi.org/10.3390/membranes12010030

Pamme, N. (2006). Magnetism and microfluidics. *Lab on a Chip*, *6*(1), 24–38. https://doi.org/10.1039/b513005k

Ranade, V. V. (2002). *Computational Flow Modeling for Chemical Reactor Engineering* (Vol. 5). Academic Press.

Ray, A. K. (1998). A new photocatalytic reactor for destruction of toxic water pollutants by advanced oxidation process. *Catalysis Today*, *44*(1–4), 357–368.

Rodriguez, P., Meille, V., Pallier, S., & Ali Al Sawah, M. (2009). Deposition and characterisation of TiO2 coatings on various supports for structured (photo)catalytic reactors. *Applied Catalysis A: General*, *360*(2), 154–162. https://doi.org/10.1016/j.apcata.2009.03.013

Safe Drinking Water Foundation. (2021). *Conventional water treatment: coagulation and filtration — Safe Drinking Water Foundation*. www.safewater.org/fact-sheets-1/2017/1/23/conventional-water-treatment

Samstag, R. W., Ducoste, J. J., Griborio, A., Nopens, I., Batstone, D. J., Wicks, J. D., Saunders, S., Wicklein, E. A., Kenny, G., & Laurent, J. (2016). CFD for wastewater treatment: an overview. *Water Science and Technology*, *74*(3), 549–563. https://doi.org/10.2166/wst.2016.249

Schlick, T. (2010). *Molecular Modeling and Simulation: An Interdisciplinary Guide* (Vol. 2). Springer.

Sengupta, T. K., Kabir, M. F., & Ray, A. K. (2001). A Taylor vortex photocatalytic reactor for water purification. *Industrial & Engineering Chemistry Research*, *40*(23), 5268–5281.

She, F. H., Gao, W. M., Peng, Z., Hodgson, P. D., & Kong, L. X. (2008). Micro- and nano-characterization of membrane materials. *Journal of the Chinese Institute of Chemical Engineers*, *39*(4), 313–320. https://doi.org/10.1016/j.jcice.2008.01.008

Takht Ravanchi, M., Kaghazchi, T., & Kargari, A. (2009). Application of membrane separation processes in petrochemical industry: a review. *Desalination*, *235*(1–3), 199–244. https://doi.org/10.1016/j.desal.2007.10.042

Wang, Y., Zhe, J., Chung, B. T. F., & Dutta, P. (2008). A rapid magnetic particle driven micromixer. *Microfluidics and Nanofluidics*, *4*(5), 375–389. https://doi.org/10.1007/s10404-007-0188-x

Ward, K., & Fan, Z. H. (2015). Mixing in microfluidic devices and enhancement methods. *Journal of Micromechanics and Microengineering*, *25*(9), 094001. https://doi.org/10.1088/0960-1317/25/9/094001

Wiley, D. E., & Fletcher, D. F. (2002). Computational fluid dynamics modelling of flow and permeation for pressure-driven membrane processes. *Desalination*, *145*(1–3), 183–186.

Wols, B. (2011). Computational fluid dynamics in drinking-water treatment. *Water Intelligence Online*, *10*, 9781780401003. https://doi.org/10.2166/9781780401003

Yadollahpour, A., Rashidi, S., Ghotbeddin, Z., Jalilifar, M., & Rezaee, Z. (2014). Electromagnetic fields for the treatments of wastewater: a review of applications and future opportunities. *Journal of Pure and Applied Microbiology*, *8*(5), 3711–3719.

2 Molecular Dynamics for the Membrane Process

Abhishek Kumar[1] and Krunal M. Gangawane[1,2*]
[1]Department of Chemical Engineering, National Institute of Technology Rourkela,
Rourkela, Odisha, India
[2]Department of Chemical Engineering, Indian Institute of Technology Jodhpur,
Jodhpur, Rajasthan, India
*Corresponding Author

CONTENTS

2.1   INTRODUCTION

As a result of growing concerns regarding water supply, researchers are actively seeking more effective methods of acquiring hygienic and healthy water for consumption and preserving the purity of natural sources of freshwater without incurring more environmental and energy problems. As a direct reaction to these concerns, technologies for the treatment of water based on membranes have emerged as a viable option in recent decades due to their capacity to enhance the efficiency and productivity of water treatment. Membrane technologies are also easy to operate and scale up for usage in many commercial processes, which can achieve excellent energy efficiency (Ebro, Kim, and Kim, 2013).

A membrane is a layer that divides two phases by limiting passage across it selectively. Membranes have existed since the beginning of the 18th century. Since then, advances have made membranes superior for many applications ('Membrane and Desalination Technologies,' 2008; Ravanchi et al., 2009).

The substance that composes a membrane might be either organic or inorganic. Polymers made from synthetic organic compounds are the building blocks used in producing organic membranes. Membranes used in pressure-driven separation processes, such as microfiltration (MF), ultrafiltration (UF), nanofiltration (NF), and reverse osmosis (RO), are often fabricated from synthetic organic polymers. Polyethene (PE), polytetrafluoroethylene (PTFE), polypropylene, and cellulose acetate are some of the other materials that fall into the organic polymers' category (Aliyu, Rathilal, and Isa 2018). Ceramics, metals, zeolites, and silica are examples of the kinds of inorganic materials that may be used to make membranes. They are chemically and thermally stable, and some of the applications that use them in the industry include hydrogen separation, ultrafiltration, and microfiltration (Baker, 2012; Mallada and Menéndez, 2008).

A wide variety of arrangements of these pressure-driven membrane processes have been implemented in a variety of different wastewater treatment systems. In some circumstances, pressure-driven membrane processes play the role of treatment for subsequent unit processes in the wastewater (Šostar-Turk, Simonič, and Petrinić, 2005). These pressure-driven membrane processes can eliminate contaminants such as color, total dissolved solids, chemical oxygen demand, urea, sodium alginate, potassium, reactive dye, and oxidizing agents (Mohammadi and Esmaeelifar, 2004; Samaei, Gato-Trinidad, and Altaee 2018), as shown in Table 2.1; some applications of these pressure-driven membrane technologies that have been carried out in WWT.

MF, UF, and NF are often used as RO pretreatment stages. It reduces membrane fouling and improves flow. It removes pollutants from wastewater using several barriers. Pressure-driven membranes have made wastewater reclamation possible. Pressure-related energy needs remain an issue (Suwaileh et al., 2018; Ezugbe and Rathilal, 2020). Since membrane technology is used more in wastewater treatment processes, we need to learn more about membranes made of synthetic materials and various separation processes. One way to reach this goal is to use a molecular dynamics (MD) computational technique.

The numerical solution to the Newtonian equations of motion may be found via MD's step-by-step computational methods. By determining equilibrium and dynamic characteristics that are generally difficult to measure using the simple analytical or popular methodology, it needs to investigate the molecular level of physical processes and mobility in systems often on a scale measured in nanometers. It is done to understand better how these things occur. As a result of these factors, in the field of research dealing with membranes, MD is becoming more important, which is based on the molecular motions of the components of membranes. Next, the following sections will demonstrate how molecular dynamics in membranes (MDM) may be used in various study fields connected to membrane-based water treatment (Schlick, 2010).

TABLE 2.1

A Few Uses for Pressure-Driven Membrane Technologies in WWT

S. No.	Pressure-Driven Membrane Technologies	Wastewater Treatment	Results	Sources
1.	Microfiltration (MF)	(a) Municipal wastewater (b) Oil sludge emulsified with synthetics	(a) Pollutants are reduced to levels under the threshold for detection. (b) 95% removal efficiency of organic compounds	(Dittrich et al., 1996) (Wang et al., 2009)
2.	Ultrafiltration (UF)	(a) Vegetable oil factory (b) Poultry slaughterhouse	(a) UF results in a decrease in COD that is 91% lower, TOC that is 87% lower, TSS that is 100% lower, $[PO^{4-3}]$ that is 85% lower, and $[Cl^-]$ that is 40% lower, respectively. (b) Over 94% of COD and BOD5 were eliminated.	(Mohammadi and Esmaeelifar, 2004) (Yordanov, 2010)
3.	Nanofiltration (NF)	(a) Textile (b) Pulp and paper mill	(a) 57% of total chemical oxygen demand, 100% of total color, and 30% of total salinity were retained. (b) Removal of total hardness and sulfate.	(Ellouze, Tahri, and Amar, 2012) (Beril Gönder, Arayici, and Barlas, 2011)
4.	Reverse osmosis (RO)	(a) Domestic wastewater treatment (b) Advanced treatment of coal chemical wastewater	(a) 100% phosphorus and 90% nitrogen were removed. (b) Biofouling caused 40% of the first-stage flow reduction and 27% of the second.	(Giraldo Mejía et al., 2022) (Sun et al., 2022)

It is essential for those who utilize the technology to get a more in-depth grasp of the concepts underlying membrane operations to fully capitalize on the benefits that membrane processes offer. In order to fully comprehend the MF and NF theories, it is necessary to do a comprehensive assessment of membrane materials and procedures (Jiang et al., 2019). Two goals that will contribute to an improvement in our knowledge of membrane processes if we make use of MDMs:

(1) The identification of the quantities that will be used to characterize the flow of particles on and across membranes.
(2) The analysis of the components that comprise the membranes and how they are characterized.

In this chapter, we will discuss the use of MDM as a method for monitoring and understanding events that occur in membrane-based water treatment systems. The possible improvements that MDM might make to our comprehension of membrane processes include providing a concise introduction to classical MDand providing a summary of the static and dynamic features that MD is able to compute. The chapter then moves on to validate the possible improvements that MDM may make to our knowledge of membrane processes. The reader should expect that MDM will be able to offer novel and beneficial information in the area of membranes and procedures for WWT.

2.2 THE FUNDAMENTALS AND OPERATIONAL PROCEDURES OF MOLECULAR DYNAMICS

2.2.1 THE FUNDAMENTALS OF MOLECULAR DYNAMICS

MD is an effective computation technique used for simulating the motion of the atom and molecule over time. There are benefits to using MD simulations rather than other computer simulation approaches because, by taking averages over time, they can figure out the system's dynamics, like transport coefficients, fluctuations that depend on time, responses to changes in the system, rheological properties, and spectra. MD figures out these numbers by combining ideas from math, chemistry, physics, and computer science. It contributes to illuminating more applied fields of study, such as material science, biology, environmental science, and nanotechnology (Attig et al., 2004) (She et al., 2008).

The fundamental idea that underpins the MD is derived from Newton's equation of motion, as well as the interactions that take place between atoms. The equation that describes the connection between the mass (m), acceleration (a), and force (F) imposed on particle i follows Eq. (2.1). The gradient may also be used to describe the Newtonian force of potential energy (U), which is demonstrated in Eq. (2.2), where F_{i}, denotes the force that is being exerted on particle i by N^{-1} other molecules, and r and t represent the distance that separates the particle time, respectively,

$$F_i = m_i a_i = m_i \frac{d^2 r_i}{dt^2} \qquad (2.1)$$

$$F_i = -\nabla_i U(r)_i = -\frac{\partial U(r^N)}{\partial r_i} \qquad (2.2)$$

Using the relationship described in Eqs. (2.1) and (2.2), it is possible to determine the trajectory of all atoms as it is characterized by their subsequent locations, velocities, and momenta. Its trajectory is then transformed into the raw data utilized for anticipating a system's bulk characteristics and further correlating those qualities to physical events.

Within a given system, pairs of atoms interact with one another regardless of whether they are linked or physically distant; these interactions are the central focus of a MD simulation. As a consequence of both bound and substantial intermolecular forces, it is possible to rearrange the equations describing potential energy such that they explain whether components in the system stretch, vibrate, and spin around the bonds, as shown in Eq. (2.3). To explain how the particles in the system stretch, vibrate, and spin around the bonds, potential energy formulas may be put up in such a way as to characterize these behaviors.

$$U_{total} = U_{bond} + U_{angle} + U_{dihedral} + U_{vdW} + U_{Coulomb} \qquad (2.3)$$

The stretching, bending, and twisting in bonded atoms are accounted for by the U_{bond}, U_{angle}, and $U_{dihedral}$ crystal structures. On the other hand, U_{vdW} and $U_{Coulomb}$ depict interactions that do not involve a bond. The van der Waal interactions, often known as U_{vdW}, are caused by the weak forces between atoms that are not linked to one another. The ionic bonds brought forth by atom charges may be measured using the $U_{Coulomb}$ system.

In the last step of the equation for determining the total potential energy, the external forces operating on the system are considered, denoted by the symbol $U_{external}$. Some examples of these forces are applied pressure and the influence of container walls. Values may be derived from the pressure that is exerted on the system as a whole throughout the processes being studied when it comes to pressure-driven membrane processes (Starr et al., 1999) (Kumar and Maiti, 2011) (Kumar, Brar, and Cledon, 2022).

2.2.2 OPERATIONAL PROCEDURES OF MOLECULAR DYNAMICS

The first step in conducting a MD simulation is to prepare the system's initial structure of the system. It can be done by using experimental data, X-ray crystallography, nuclear magnetic resonance spectroscopy, or other spectroscopic techniques, using computational approaches, such as homology modeling. The structure must be energy-minimized to remove any bad contacts or steric clashes between atoms. The next step is to prepare the simulation box. The simulation box is a rectangular box that contains the system and the solvent. The dimensions of the simulation box and the type of solvent that can be used is water, but other solvents such as methanol or dimethyl sulfoxide (DMSO) can also be used. After preparing the simulation box, the system will then be equilibrated. For this purpose, the temperature and pressure will need to be raised incrementally until the system achieves a state of equilibrium. Depending on the system, the time it takes for the equilibration process may range anywhere from several hundred picoseconds to several nanoseconds. The last thing to be done is to start the manufacturing run. The real simulation is the production run, during which the system is developed over time. The simulation results may be examined using various methods, such as the root mean squared deviation (RMSD) or the radius of gyration (Mótyán, Tóth, and Tőzsér 2013) (Seyyedattar, Zendehboudi, and Butt 2019).

2.2.3 Various Approaches to Modeling and Simulation in MD

There are two main approaches in MD research on water treatment systems through membrane-based material: equilibrium simulations and nonequilibrium molecular dynamics (NEMD). As Ebro, Kim, and Kim (2013) explained, the primary alteration among simulations of equilibrium and those of nonequilibrium conditions. In addition, they demonstrated that a comprehensive analysis of the characteristics of behavior in small channels might be achieved by doing research using both of these approaches. It was one of their main contributions to the field. They emphasized in their paper that under equilibrium circumstances, there should not be any variation in the difference in chemical potential between the two solution reservoirs that are kept apart by a conduit. It is important to keep it in mind. In contrast, using a simulation that did not include equilibrium, the researchers applied a constant impact on a layer composed of water molecules that were a given length thick to determine the change in hydrostatic pressure (Jatkar, Lee, and Lee 2010).

Equations of equilibrium may be helpful for various research, including those in which computations of static and dynamic property values are performed. However, NEMD has been demonstrated to be more effective than equilibrium MD when calculating transport coefficients. Although equilibrium MD may also be used to construct temporal correlation functions, it is not required. Numerous research has used it to calculate shear stress as well as bulk viscosity, coefficient of thermal expansion, and diffusion coefficients among the properties that may be measured. (Haile, 1992; Sharma and Kumar, 2019). In addition, it has been suggested that one of the many benefits of using MD is the possibility of using it to investigate a system that has deviated from its equilibrium as an experimentation tool. NEMD is a significant tool for understanding actual phenomena for membrane-based water treatment operations. It is especially true when considering that one of the key research focuses in this area is the diffusion characteristic of membranes. NEMD was used to calculate the diffusion coefficients of various simulation systems and kinds of membranes. Various MD approaches have been devised to carry out NEMD. The dual control volume grand canonical molecular dynamics (DCV-GCMD) model is an excellent illustration. This methodology, which may be seen as a hybrid GC Monte Carlo/MD method, enables the management of potential chemical gradients inside the simulation system of necessary quality for membrane-based methods of water treatment (Heffelfinger and van Swol, 1994) (Thompson and Heffelfinger, 1999).

2.3 THE EXECUTION OF SIMULATIONS OF MOLECULAR DYNAMICS

When doing research, employing computer simulations of molecular dynamics requires dealing with particle trajectories. As seen in Figure 2.1, the steps involved dynamic molecular modeling, which can further be divided into two steps: develop model and MD simulation; the primary emphasis of MD simulation is on the generation of trajectories, which are subsequently subjected to an analysis to characterize the attributes of the simulated system.

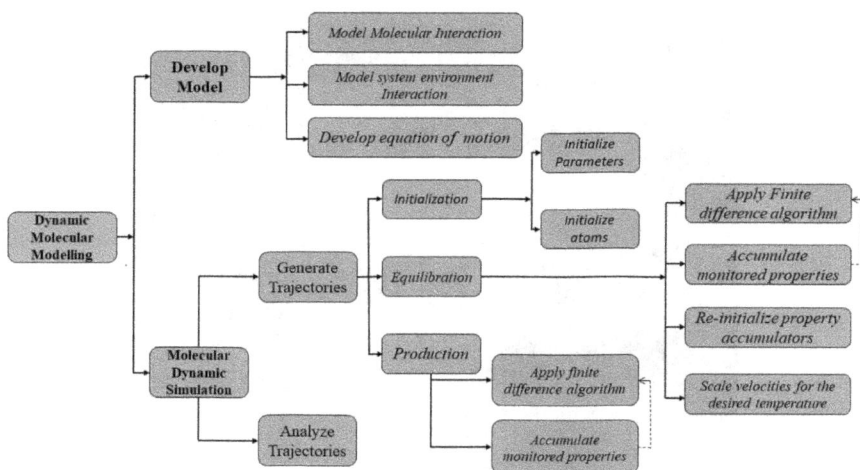

FIGURE 2.1 Steps involved for the dynamic molecular modeling, dotted lines represent iterative stages (Haile et al. 1993).

The production of the trajectories is an essential part of MD simulations since it produces the "raw data" that then has to be processed and evaluated to acquire the parameters of the membrane process that is being simulated. The generation of phase-space trajectories follows a basic flow that is made up of three stages: the initialization stage, the equilibration stage, and the production stage. This flow is adhered to throughout the process (Haile et al., 1993) (Ebro, Kim, and Kim 2013).

During the initialization process, the modeler will set up the prerequisites (such as system units, the quantity and composition of the particles, the size of the system, boundary conditions, and other simulation characteristics, etc.) as well as initialize the atoms (i.e., set initial position and velocity). Two distinct simulation configurations are presented here, as seen in Figure 2.2: (a) the configuration for a simulation of forward osmosis, which consists of two compartments with varying salt concentrations that are separated by two layers of carbon nanotube (CNT) membranes (Jia et al., 2010), and (b) a configuration for running a model of water propagation via CNTs (Zhu and Schulten, 2003).

In the equilibration process, a real MD run is performed to let the amounts inside the system settle down to their equilibrium averages. It is the solvent that relaxes with the solute as the system stabilizes. For instance, developing a water treatment method based on membranes requires that water molecules should be allowed enough time to equilibrate with the ions, foulant, and membrane particles. It ensures that the model accurately represents the real-world process. It is essential to do so to guarantee precise outcomes. The equilibration process assures that the system will have no recollection of the initial circumstances pre-allocated to it or those chosen randomly. It is another safeguard taken to guarantee that the qualities that will be tested will not be altered in any way. The portion of a water flow model known as the equilibration

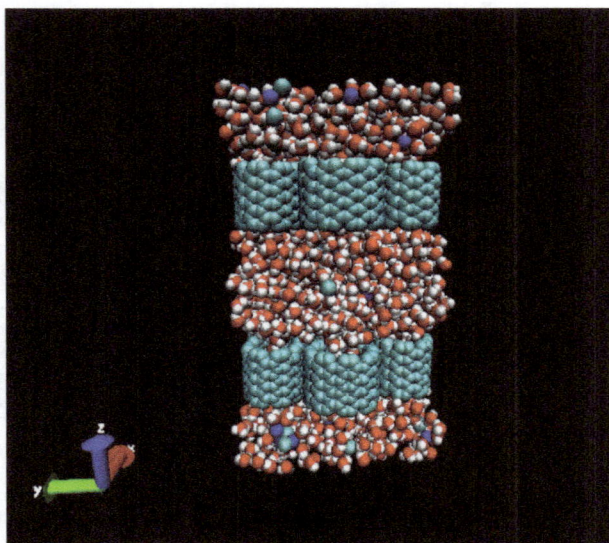

FIGURE 2.2(a) FO desalination simulation setup. Low concentration (center) and high concentration (ends, joined by periodic boundary conditions) CNT layers divide NaCl solution chambers. Adopted by Jia et al. (2010).

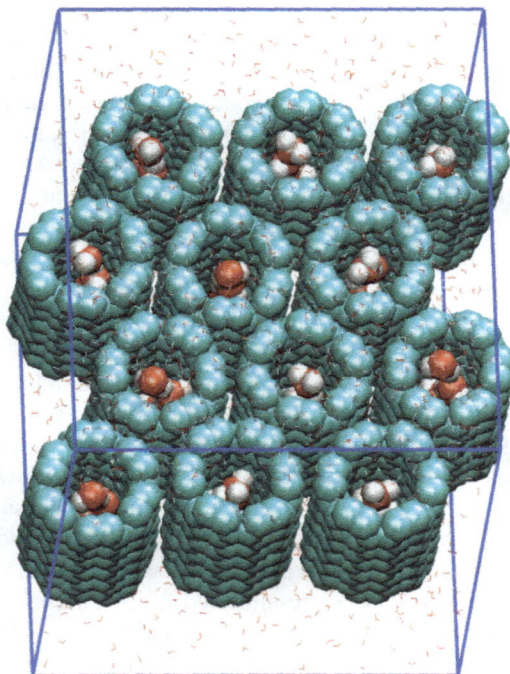

FIGURE 2.2(b) Zhu and Schulten (2003) replicate the unit cell of the system. In the simulation box with 6348 atoms, 12 hexagonally packed (6,6) CNTs are displayed.

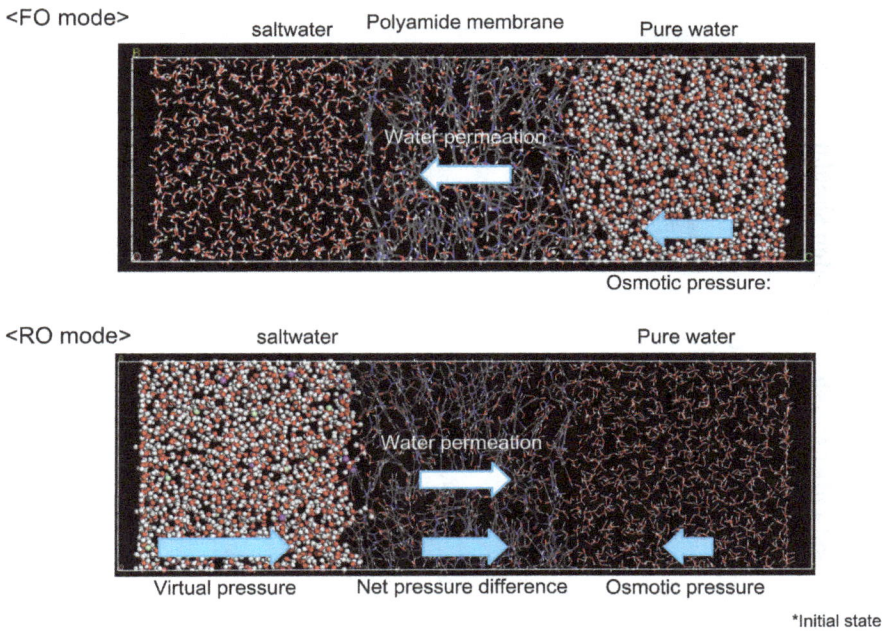

FIGURE 2.3 A polyamide RO membrane simulation's equilibration phase. (Yoshioka et al., 2018).

phase and the level of salt that may pass through a polyamide RO membrane is shown as having finished in Figure 2.3.

After everything has settled down and the system has achieved equilibrium, it is finally time to begin the phase of data collecting. During the stage of the manufacturing process, a specified number of time steps are used to run the MD simulation. As shown in Figure 2.1, in the production phase, researchers also make advantage of the equilibration process's first and second phases; these stages are intended to create the raw data required for further analyses and predictions of the system's features (Yoshioka et al., 2018).

2.4 DEVELOPMENT OF MEMBRANE-BASED WATER TREATMENT USING MOLECULAR DYNAMICS

The behavior of water and other molecules' interactions with membrane surfaces may be studied using MD simulations, which are used in water treatment processes. MD simulations can also help researchers understand how different membrane materials and surface properties affect the separation of ions and molecules and can also be used to optimize membrane design and performance. Some studies have used MD simulations to investigate the transport of ions, such as sodium and chloride, through a membrane. In contrast, others have focused on removing specific contaminants, such as heavy metals or organic pollutants, as listed in Table 2.2. Overall, MD simulations

TABLE 2.2
Molecular Dynamic Membrane Activities Associated with Water Treatment and Other Membrane-Related Simulations

S. No.	Source	Subject of Study	Membrane Type	Other Components	Results Observed/Quantities Measured
1.	(Uchida and Matsuoka, 2004)	A variety of aqueous salt solutions, each with a unique concentration and set of ionic characteristics	Without membrane	Sodium chloride, H_2O	• Radial distribution function (RDF) • Coordination number • Osmotic pressure
2.	(Mao and Zhang, 2012)	Different types of water and their defining characteristics	Without membrane	Water	• Shear viscosity • specific heat of rigid water models • Thermal conductivity
3.	(Murad, 1996)	The processes of osmosis and reverse osmosis were performed at varying concentrations.	Semi-permeable membranes	Solvent and solute models	• Density profiles of both the solute and solvent molecules • Osmotic pressure
4.	(Sohail Murad and Nitsche, 2004)	The influence of the pore size, structure, and diameter of the RO membrane on the flow and selection	Harmonically linked barrier molecules.	Sodium chloride, H_2O	• Solvent flux • Ion selectivity
5.	(Luo et al., 2011a)	A polymeric membrane's water flow and saline rejection characteristics	FT-30 polymerized membrane	Sodium chloride, H_2O	• Diffusion of water and ions • Solvation of ions • Free energy
6.	(Harder et al., 2009)	Polymerization, as well as the production of membranes	Polymerized membrane	Sodium chloride, H_2O three types of MPD polyamides, TMC	• Membrane polymerization • Water flux • Salt rejection • Free energy

No.	Reference	Description	Material	Medium	Parameters
7.	(Suk, Raghunathan, and Aluru, 2008),(Hughes et al., 2011)	RO is constructed from a variety of materials, including carbon nanotubes (CNT), carbon nanotubes (BNNT), polymethyl methacrylate (PMMA), and zeolite	CNT pore; BNNT pore; PMMA membrane; zeolite membrane	Sodium chloride, H_2O model ions	• Potential of mean force (PMF) • Water diffusion coefficient • RDF • Salt behavior
8.	(Pascal, Goddard, and Jung, 2011), (Wang, Dumont, and Dickson, 2012), (Kalra, Garde, and Hummer, 2003)	The flow of water over carbon nanotubes Flow of water over carbon nanotubes	CNT pore	H_2O	• Function of the axis of distribution • Free energy, entropy, enthalpy • Density-velocity pattern • Water occupancy/conduction • Energy for water binding
9.	(Goldsmith and Martens, 2010)	Ion rejection and separation utilizing charged nanotubes; ion and water transport in CNTs for field-effect transistors	CNT pore; modified CNT pore	Sodium chloride, H_2O	• Permeability • Ion rejection and separation • Free energy profile • Orientation of the water • Adjustments to the surface charge

in membrane-based water treatment research can provide valuable insights into the underlying physical and chemical processes that govern membrane performance and aid in developing more efficient and effective water treatment systems.

2.4.1 MOLECULAR DYNAMICS TO DESIGN MEMBRANE MATERIALS FOR WATER TREATMENT PROCEDURES

MD simulations can be used to build and modify the structures of the water treatment method based on membranes required and investigation of their physical and surface properties. MD simulations involve computer algorithms to simulate a system's interactions between atoms and molecules. It is able to provide very specific information on the structure and behavior of the membrane material, such as its permeability, mechanical properties, and stability. Information can be useful for developing new membrane materials and optimizing existing ones for specific water treatment applications (Hollingsworth and Dror, 2018).

The structures of the polyamide skin layer of a RO membrane can be seen using MD and other computer simulation methods. In the study, Hirose (1997) showed that to make high-performance composite RO membranes in the future; researchers would have to look into the molecular design of the cross-linked polyamide skin layers of membranes. It would require molecular simulations and calculations.

In another study (Harder et al., 2009), it was possible to show how a polyamide membrane is made by simulating the polymerization of its monomers, as shown in Figure 2.4. There were four steps to the method, starting with stage 0. 250 m-phenylene diamine (MPD) and benzene 1,3,5-tricarboxylic acid chloride (or trimethyl chloride, TMC) monomers were placed randomly, and MD runs were used to simulate polymerization until the target membrane density was reached. In the first stage, amide bonds are made continuously until the rate of change of polymerization is almost zero. The amine and acyl chloride groups were linked by adding an intermolecular potential in the second stage. The last stage starts when the rate of change of bond formation over time is close to zero. In stage 3, free acyl chloride monomers were moved close to unreacted amine groups, and the remaining free monomers were removed after simulation cycles.

A very limited number of studies have been published that discuss MD simulations of polymeric RO membranes. One possible explanation for the challenging nature of such simulations is that simulating membranes like FT-30 is more difficult than other membranes because it requires more complex molecules (initial monomers with complex structures). Most importantly, the simulation's polymerization process includes breaking and forming many bonds. Other membranes, such as CNT and BNNT, are much easier to model than FT-30. In addition, to carry out these simulations, you will need much more computational power, longer simulation time frames, and bigger systems (with a higher total number of atoms). Nevertheless, doing MD simulations of polymeric RO membranes would open up opportunities to understand the properties and the behavior of these types of membranes, as well as to improve currently available membranes and develop new and improved ones. Despite the difficulties associated with using this method (Luo et al., 2011a, 2011b), it is true.

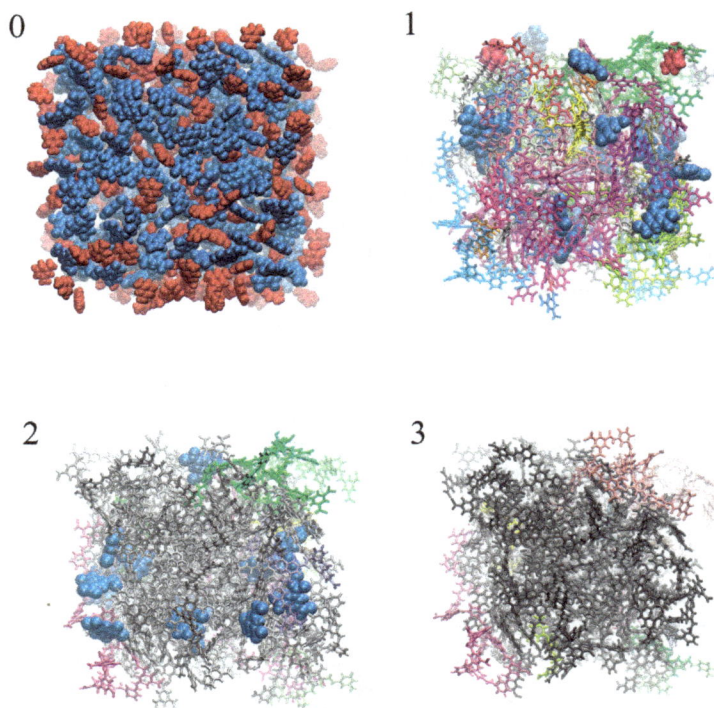

FIGURE 2.4   The development of the membrane proceeds via these four stages (Harder et al., 2009).

Much research has been conducted using MD simulations to study the consequences of employing nanotechnology to strengthen membranes. CNTs and boron nitride nanotubes (BNNTs) are two nanomaterials of interest because of their high ion rejection and water flow rates (Hilder, Gordon, and Chung, 2009). Recent research has shown that nanotubes are gaining appeal in membrane-based water treatment technologies because of their high efficiency. It can design novel characteristics and functions for these nanoparticles once these nanomaterials' internal and surface chemistry, atomic structure, and assembly are controlled (Riviere and Myhra, 2009). As seen in Table 2.2, several simulations have been carried out to develop nanotubes for water treatment procedures and gain knowledge of their qualities. When investigating the selectivity and permeability of membranes, it is helpful to understand the membrane's pore size, thickness, and porosity, among other physical elements of the membrane's features. According to the research (Parvatiyar, 1998), it is possible to reduce the amount of entropy created in the fluid during the filtering process by adjusting factors such as the diameter of the membrane tube, the Reynolds number, and the flow conditions. Because of his efforts, he could observe the connection between the development of entropy and the concentration polarization that occurs during UF. In addition, Hilder reported the radial density distribution of water molecules (Hilder, Gordon, and Chung, 2009).

2.4.2 Water and Ion Transport Across the Membrane

In membrane-based water treatment operations, one of the most important factors to take into account is how easily water and ions may pass through the membrane. This movement, also known as transport, is affected by the characteristics of the membrane, which include the charge in porosity of the membrane, as well as the properties of the water and ions themselves, such as their size and charge. Proper design and operation of the membrane system can maximize the transport of desired species while minimizing the transport of unwanted species. The computation of ion and water fluxes is possible using MD simulations. Table 2.3 compares the water diffusion coefficients recorded in several studies analyzed for this article. These coefficients were measured both inside the membrane and in the bulk solution. The research included in this table attempted to model a polyamide RO membrane with an experimental density of 1.38 g/cm^3 (Mi, Cahill, and Mariñas, 2007). The diffusion coefficients were not affected by the fact that the membranes were manufactured utilizing various approaches to their construction. Because of this, it can be deduced that the simulation approach for producing the membranes has a negligible impact on the water diffusion coefficient estimated inside the membrane.

Furthermore, the concentration of sodium ions in the solution phase does not significantly impact the water's ability to diffuse across the membrane, as stated by Hughes and Gale (2010). These findings are based on research conducted. Therefore, despite the simulations in Table 2.3 being carried out with various amounts of salt, D_m values have remained constant and have maintained the same general scale of measurement. Regarding the water vapor diffusion coefficient in the bulk phase of the atmosphere, it is reasonable to anticipate that different works will give varying results. This is because different works will use a different water model, and they will also use different simulation parameters (such as cutoffs). MD simulations may also be used in the investigation of osmosis and osmotic permeability, which is closely related to the problem of calculating the values of the diffusion coefficient. In their paper, Luo and Roux (2010) developed a universal approach for determining the

TABLE 2.3
Water Diffusion Coefficients in Solution and Cross-Linked Polymer Membranes Computed Using Various MD Solutions

S. No.	Source	Density of Actual Membranes (g/cm^3)	Coefficient of Water Diffusion Inside the Membrane, D_m (m^2/s)	At 27 °C, the water diffusion coefficient in a bulk solution, D_w (m^2/s)
1.	(Kotelyanskii, Wagner, and Paulaitis, 1998)	1.38–1.40	0.20×10^{-9}	2.45–8.02×10^{-9}
2.	(Yeap et al., 2017)	1.4	0.50×10^{-9}	5.1×10^{-9}
3.	(Hughes and Gale, 2010)	1.34	0.21×10^{-9}	2.46×10^{-9}
4.	(Luo et al., 2011b)	1.34	0.50×10^{-9}	Not specified

osmotic pressure of ionic solutions. Their simulation setup included a virtual wall that exerted an instantaneous force on the system. In their study, Zhu et al. (Zhu, Tajkhorshid, and Schulten, 2001) compared and contrasted a material's osmotic and diffusion permeability. Computing the osmotic passage of water in settings of equilibrium and nonequilibrium required them to construct a "collective diffusion model," which they did in a separate piece of work. In 1996 (Murad, 1996), the last in a series of MD simulations that explicitly investigated osmosis and reverse osmosis was finished. In this particular piece of research, the transition from FO to RO was investigated by adjusting either the concentration of the solution or the pressure differential across the membrane. In order to investigate the effects of adsorption along the walls, density profiles of the solute and the solvent were obtained.

The fundamentals of thermodynamics have the potential to provide transport characteristics that characterize a membrane process. It is possible to determine the significant quantities related to thermodynamics using MD simulations. These quantities include energy and pressure and can be used to calculate free energy differences (Hwang, 2004). When these differences in free energy are understood, it should be feasible to get insight into the driving force behind a process, and it would also make it possible to produce a "direct and quantitative comparison" based on the findings of experiments (De Jong et al., 2011). The behavior at the membrane contact may also be described using free energy profiles. Thermodynamic parameters may be used to describe the wetting phenomena that occur within a nanotube. It is possible to describe the flow of water within CNTs of varying diameters by referring to the free energy, entropy, and enthalpy of the system. The MDM technique may also investigate membrane processes influenced by electric fields. In the research carried out by Murad (Murad, 2011), MD simulations were used to investigate the role that external fields have in determining the dynamic behavior of polymer electrolytes. In particular, they focused their attention on investigating parameter factors, including ion mobility, drift velocity, and drift-diffusion rate, all of which have the potential to influence membrane processes such as ED and EDI. Additionally, Kim and Darve (2009) explored electroosmotic flows in nanochannels and discovered the effects of an electric field on water molecules.

2.4.3 Minimizing Membrane Fouling

The elimination of excessive membrane fouling is one of the numerous obstacles that need to be overcome in membrane-based water treatment procedures. Nevertheless, high operating expenses and poor product production due to fouling are unavoidable (Kim and Darve, 2009; Moonkhum et al., 2010). According to Chen et al. (Chen, Li, and Fane, 2004), thermodynamics of synthetic membrane materials and membrane-fluid boundary layer-bulk fluid interactions may increase filtering efficiency and fouling avoidance. Thermodynamics of natural membrane materials may also be useful. In forthcoming research on membrane fouling, the identification and characterization of membrane foulants should be included, describing their interactions, and developing procedures for visualizing the foulants. In light of these suggestions, MD simulations have previously been carried out to explain the fouling phenomena in membranes.

Simulations were performed by Hughes and Gale (2010) to investigate how possible foulants, including glucose, phenol, and oxygen, interacted with RO membranes. They also showed in a prior study that MD simulations, with the assistance of a graphics tool, may be used to visualize fouling layers on a membrane and give additional information. It was accomplished via the use of MD simulations. It is anticipated that in the future, MD simulations conducted to examine fouling will include studies regarding the fouling process and more complicated foulants. By measuring the intermolecular adhesion forces, atomic force microscopy, often known as AFM, is commonly used to evaluate fouling in membranes. There have been research on the use of MD simulations to further evaluate and understand AFM findings, indicating that there is a high likelihood of correlating the results of MD simulations to the AFM data collected from membrane processes. Despite the fact that these studies did not particularly address membrane-based water treatment, they did mention the use of MD simulations.

2.5 APPLICATIONS IN ELITE FIELDS

Molecular dynamics simulations have a diverse set of possible uses and applications in the study of biological membranes, including the following:

1. Lipid-protein interactions:
    Lipid-protein interactions are a critical part in the formation and operation of biological membranes, both in terms of structure and function, and having an awareness of these interconnections is crucial for comprehension of the dynamics of these membranes. Molecular dynamics simulations are a powerful tool that can be used to study lipid-protein interactions and gain insights into biological membranes' dynamics. Molecular dynamics simulations involve computer simulations that model interactions between individual atoms or molecules. These simulations may be used to examine lipids and proteins in a biological membrane and comprehend their dynamics. One of the key advantages of using molecular dynamics simulations to investigate lipid-protein interactions is that they can provide a high level of detail about the interactions between individual atoms and molecules. Its level of detail can be used to gain insights into the specific interactions between lipids and proteins. It can help to understand how these interactions contribute to the overall dynamics of the membrane (Pitman et al., 2005).

    Moreover, MD simulations can be used to study the effects of different factors on lipid-protein interactions. For example, simulations can be run to study the effects of changes in temperature or pH on the interactions between lipids and proteins. It can provide valuable information about how these interactions are affected by environmental changes. Another advantage of using molecular dynamics simulations to study lipid-protein interactions is that they can be used to study the behavior of large systems, such as biological membranes. It can provide valuable insights into the overall dynamics of these systems. It can help to understand how the interactions between lipids and proteins contribute to the overall behavior of the membrane. Overall, molecular dynamics simulations are a powerful tool for studying lipid-protein interactions in biological membranes. They can provide a high level of detail about the interactions between individual atoms and molecules and can be used to study the effects of different factors on these interactions. Additionally, they can be used to study the behavior of large systems, such as biological membranes, and can provide valuable insights into the overall dynamics of these systems (Hollingsworth and Dror, 2018) (Pitman et al., 2005).

2. Membrane protein structure and function:
    MD simulations can study the structure and function of membrane proteins, including their dynamics, interactions, and conformational changes. Membrane proteins are a critical component of biological membranes, playing a vital role in the transport of molecules across the membrane, as well as in the regulation of cell signaling and communication. Understanding the structure and function of these proteins is essential for understanding the dynamics of biological membranes, and molecular dynamics simulations are a powerful tool that can be used to study these proteins and gain insights into their behavior. Membrane proteins are unique because they are embedded in the membrane's lipid bilayer (Zhu, Tajkhorshid, and Schulten, 2001), which challenges understanding their structure and function. Traditional techniques such as X-ray crystallography and NMR spectroscopy are not always effective in studying these proteins due to their inherent flexibility and dynamic nature. Molecular dynamics simulations, on the other hand, can provide a detailed understanding of the interactions between the protein and the surrounding lipids and can help to understand how these interactions contribute to the overall dynamics of the membrane. Molecular dynamics simulations involve computer simulations that model interactions between individual atoms or molecules. These simulations can be used to study the behavior of membrane proteins in different environments. They can provide a high level of detail about the interactions between individual atoms and molecules. The level of detail can be used to understand the specific interactions between the protein and the surrounding lipids, such as the formation of specific binding sites and the effect of these interactions on the stability and function of the protein (Srivastava et al., 2018) (Ivetac and Sansom, 2008). In addition, molecular dynamics simulations can be used to study the effects of different factors on the structure and function of membrane proteins. For example, simulations can be run to study the effects of changes in temperature, pH, or the presence of ligands on the structure and function of a membrane protein. It can provide valuable information about how these proteins are affected by environmental changes and help understand the mechanisms that underlie these changes. Another advantage of using molecular dynamics simulations to study membrane protein structure and function is that they can be used to study large systems, such as a biological membrane, allowing us to understand the overall dynamics of these systems (Khalili-araghi et al., 2010). Overall, molecular dynamics simulations are a powerful tool for studying the structure and function of membrane proteins in biological membranes. They can provide a high level of detail about the interactions between individual atoms and

molecules and can be used to study the effects of different factors on these proteins. Additionally, they can be used to study large systems, such as biological membranes, and provide valuable insights into the overall dynamics of these systems.

3.   Drug-membrane interactions:

MD simulations can be used to study the interactions between drugs and lipids in the membrane, including the binding of drugs to specific proteins and the effects of drugs on the structure and function of the membrane. Drug-membrane interactions are a crucial aspect of understanding the behavior of drugs within living cells. Studying these interactions can provide valuable insights into how drugs interact with cell membranes and how they are transported across them. MD simulations are a powerful tool for studying drug-membrane interactions. These simulations involve using computational models to simulate the behavior of individual atoms and molecules in a system, allowing researchers to study the complex dynamics of drug-membrane interactions at the molecular level (Róg, Girych, and Bunker, 2021). MD simulations can study a wide range of drug-membrane interactions, including the binding of drugs to specific receptors on the membrane, the penetration of drugs across the membrane, and the effects of drugs on the structure and function of the membrane itself. One important aspect of drug-membrane interactions that can be studied using MD simulations is the binding of drugs to specific receptors on the membrane. These receptors are protein molecules embedded in the membrane and act as gatekeepers, controlling the movement of molecules across the membrane. MD simulations can study the binding of drugs to these receptors, providing insights into how the drugs interact with the receptors and how they are transported across the membrane. Information can be used to design new drugs that have improved binding affinity and specificity for these receptors, leading to more effective treatments for many diseases. Another important aspect of drug-membrane interactions that can be studied using MD simulations is the penetration of drugs across the membrane. It is known as transmembrane transport, a critical step in the action of many drugs. MD simulations can study the transport of drugs across the membrane, providing insights into the mechanisms by which drugs are transported and the factors that affect their transport. Information can be used to design new drugs that have improved transmembrane transport, leading to more effective treatments for various diseases (Saurabh et al., 2020) (Kopeč, Telenius, and Khandelia, 2013).Overall, MD simulations are a powerful tool for studying drug-membrane interactions, providing valuable insights into the behavior of drugs within living cells. These simulations can be used to design new drugs that have improved binding affinity, specificity, and transmembrane transport, leading to more effective treatments for a wide range of diseases.

4.   Membrane fusion and fission:

MD simulations can examine membrane fusion and fission, including fusion pore generation, dynamics, and protein roles. Endocytosis, exocytosis, and intracellular transport need membrane fusion and fission. For generating novel therapies and investigating disorders where abnormalities in these mechanisms cause aberrant cell activity, molecular understanding is essential. In MD simulations, computational fluid dynamics (CFD) has become a significant tool for investigating these phenomena. In a CFD-MD technique, coarse-grained and atomistic lipid and protein models are used to simulate membrane fusion and fission. These models mimic system dynamics using molecules' physical and chemical interactions. CFD-MD models provide hydrodynamic force studies. In biology, surrounding fluids like the cytosol exert these forces, which may affect system dynamics. CFD-MD simulations may examine how hydrodynamic forces affect fusion protein binding and conformational changes or the mechanical characteristics of the bilayer that allow it to fuse or fission. CFD-MD simulations may also analyze temperature variations in the system. These variations may dramatically alter biological system dynamics. Transient features like pores may aid fusion or fission.

CFD-MD simulations may also examine how environmental circumstances affect the system. Simulations can reveal how temperatures, pressures, and chemical concentrations impact the system's dynamics. CFD-MD simulations provide a full knowledge of membrane fusion and fission mechanisms and uncover possible therapeutic targets. CFD-MD is a strong and adaptable tool for understanding these complicated biological processes since it can examine hydrodynamic forces, thermal fluctuations, and environmental factors.

5. Lipid and protein design:
    MD simulations can create stable, drug-binding lipids and proteins. Understanding and influencing biological membrane dynamics requires designing lipids and proteins. Creating accurate and realistic system models in MD simulations requires lipid and protein design. Researchers may better understand membrane dynamics by studying hydrodynamic forces and thermal fluctuations using computational fluid dynamics (CFD) and MD simulations. MD simulations need force field selection for lipid design. Force fields are mathematical models of the atom and molecule interactions. For precise and realistic results, choose the right force field. The lipid model is also crucial to lipid design. Coarse-grained or atomistic models can model lipids. Coarse-grained models are computationally easier but may not reflect lipid behavior. Atomistic models reflect lipids more accurately but need more processing. MD simulations need protein design to provide accurate and realistic system models. Protein design involves force field selection (Róg, Girych, and Bunker, 2021) (Pourmousa and Pastor, 2018).

    For precise and realistic results, choose the right force field. Protein model selection is also crucial. Coarse-grained or atomistic models can model proteins. Coarse-grained models are computationally easier but may not reflect protein activity. Atomistic models depict proteins more accurately but need more processing. Experimental data may also enhance lipid and protein, design models. X-ray crystallography data may improve protein and lipid design. MD simulations for membrane dynamics need lipid and protein design. CFD and MD simulations enable researchers to analyze hydrodynamic forces and thermal variations in the system, improving their knowledge of membrane dynamics. Accurate and realistic system models require selecting the right force field, modeling technique, and experimental data. Drug development, biotechnology, and the pharmaceutical industry employ these simulations to study biomolecule behavior at the molecular level and build novel medications and cures (Fu, Zhao, and Chen 2018).

2.6 FUTURE MEMBRANE-BASED WATER TREATMENT OF MOLECULAR DYNAMICS MEMBRANE

It is possible to utilize MD simulations to study the behavior of water and other molecules at the atomic level. They can potentially be useful for designing and enhancing water purification methods that are based on membranes in the future. Through modeling the interactions that take place between individual water molecules and the outside of a membrane, MD can help researchers understand how different membrane materials and structures affect the separation of contaminants from water. This information can be used to develop novel membranes with enhanced performance and selectivity for various applications in water treatment. In addition to being used for well-known water treatment processes including reverse osmosis, nanofiltration, ultrafiltration, and multi-stage forward osmosis, MDM may be utilized to investigate relatively novel technologies, including such membrane distillation, forward osmosis, and pressure retarded osmosis. These processes include reverse osmosis, nanofiltration, ultrafiltration, and membrane filtration. According to the authors' best knowledge, there has not been any MDM research that has been published that is connected to membrane distillation or temperature-driven membrane processes as of this time. It is also anticipated that the findings of MDM simulations will be comparable to those of FO simulations or experiments that used polymeric membranes (Jung et al., 2011; Park et al., 2011). Additionally, MDM can potentially contribute to creating an efficient PRO membrane, which may lead to the implementation of PRO in a practical and large-scale setting (Kim, Lee, and Kim, 2012). In addition to researching the topics discussed previously, MDM may also be used to investigate the following three membrane processes in their overall context, which are listed below:

- Membrane distillation
  - The influence of temperature on processes that are thermally driven.
- Forward osmosis (FO)
  - Analysis of the parameters of the feed solution and the draw solution
  - The efficiency of FO in polymeric membranes and nanotube membranes is compared.
- Pressure-retarded osmosis (PRO)
  - Interaction of feed and draw solutions with the membrane's active and porous support layer and the implications of such an interaction on generating power.
  - The design of a membrane that uses pressure to retard osmosis.

The study subjects described in this article are only a small sample of the many feasible inquiries employing MDM. As a result the MD simulations are adaptable, provided that the appropriate inputs, computer software and hardware, and theoretical understanding are all available. Depending on the amount of time and processing power made accessible, the accuracy of these simulations may be increased. As we go further into comprehending the procedures involved in water purification, the authors are of the opinion that MDM will tend to be a highly efficient instrument in membrane research. MD simulations could also be combined with other computational tools, such as machine learning, to create predictive models that can optimize membrane-based water treatment systems based on the characteristics of the water being treated. However, it is worth noting that MD simulations are still computationally intensive and require significant computational resources, so further developments in computational capabilities may be needed to realize their potential in membrane-based water treatment fully.

REFERENCES

Aliyu, Usman Mohammed, Sudesh Rathilal, and Yusuf Makarfi Isa. 2018. "Membrane Desalination Technologies in Water Treatment: A Review." *Water Practice and Technology* 13 (4): 738–52. https://doi.org/10.2166/wpt.2018.084.
Attig, Norbert, Kurt Binder, Helmut Grubmüller, and Kurt Kremer. 2004. "Computational Soft Matter: From Synthetic Polymers to Proteins." *John von Neumann Institute for Computing (NIC), Juelich* 23.
Baker, Richard W. 2012. *Membrane Technology and Applications*. John Wiley & Sons.
Beril Gönder, Z., Semiha Arayici, and Hulusi Barlas. 2011. "Advanced Treatment of Pulp and Paper Mill Wastewater by Nanofiltration Process: Effects of Operating Conditions on Membrane Fouling." *Separation and Purification Technology* 76 (3): 292–302. https://doi.org/10.1016/j.seppur.2010.10.018.
Chen, V, H Li, and A G Fane. 2004. "Non-Invasive Observation of Synthetic Membrane Processes–a Review of Methods." *Journal of Membrane Science* 241 (1): 23–44.
Dittrich, J, R Gnirss, A Peter-Fröhlich, and F Sarfert. 1996. "Microfiltration of Municipal Wastewater for Disinfection and Advanced Phosphorus Removal." *Water Science and Technology* 34 (9): 125–31. https://doi.org/10.2166/wst.1996.0192.
Ebro, Hannah, Young Mi Kim, and Joon Ha Kim. 2013. "Molecular Dynamics Simulations in Membrane-Based Water Treatment Processes: A Systematic Overview." *Journal of Membrane Science* 438: 112–25. https://doi.org/10.1016/j.memsci.2013.03.027.
Ellouze, Emna, Nouha Tahri, and Raja Ben Amar. 2012. "Enhancement of Textile Wastewater Treatment Process Using Nanofiltration." *Desalination* 286 (February): 16–23. https://doi.org/10.1016/j.desal.2011.09.025.
Ezugbe, Elorm Obotey, and Sudesh Rathilal. 2020. "Membrane Technologies in Wastewater Treatment: A Review." *Membranes*. https://doi.org/10.3390/membranes10050089.
Fu, Yi, Ji Zhao, and Zhiguo Chen. 2018. "Insights into the Molecular Mechanisms of Protein-Ligand Interactions by Molecular Docking and Molecular Dynamics Simulation: A Case of Oligopeptide Binding Protein." *Computational and Mathematical Methods in Medicine* 2018. https://doi.org/10.1155/2018/3502514.
Giraldo Mejía, Hugo Fernando, Javiera Toledo-Alarcón, Barbara Rodriguez, José Rivas Cifuentes, Francisco Ovalle Porré, María Paz Loebel Haeger, Natalia Vicencio Ovalle, Carmen Lacoma Astudillo, and Andreina García. 2022. "Direct Recycling of Discarded Reverse Osmosis Membranes for Domestic Wastewater Treatment with a Focus on Water Reuse." *Chemical Engineering Research and Design* 184 (August): 473–87. https://doi.org/10.1016/j.cherd.2022.06.031.

Goldsmith, Jacob, and Craig C Martens. 2010. "Molecular Dynamics Simulation of Salt Rejection in Model Surface-Modified Nanopores." *The Journal of Physical Chemistry Letters* 1 (2): 528–35.

Haile, J. M., Ian Johnston, A. John Mallinckrodt, and Susan McKay. 1993. "Molecular Dynamics Simulation: Elementary Methods." *Computers in Physics* 7 (6): 625. https://doi.org/10.1063/1.4823234.

Haile, James M. 1992. *Molecular Dynamics Simulation: Elementary Methods.* John Wiley & Sons, Inc.

Harder, Edward, D. Eric Walters, Yaroslav D. Bodnar, Ron S. Faibish, and Benoît Roux. 2009. "Molecular Dynamics Study of a Polymeric Reverse Osmosis Membrane." *Journal Physical Chemistry of B* 113 (30): 10177–82. https://doi.org/10.1021/jp902715f.

Heffelfinger, Grant S., and Frank van Swol. 1994. "Diffusion in Lennard-Jones Fluids Using Dual Control Volume Grand Canonical Molecular Dynamics Simulation (DCV-GCMD)." *The Journal of Chemical Physics* 100 (10): 7548–52. https://doi.org/10.1063/1.466849.

Hilder, Tamsyn A., Daniel Gordon, and Shin Ho Chung. 2009. "Salt Rejection and Water Transport through Boron Nitride Nanotubes." *Small* 5 (19): 2183–90. https://doi.org/10.1002/smll.200900349.

Hirose, Masahiko. 1997. "The Relationship between Polymer Molecular Structure of RO Membrane Skin Layers and Their RO Performances." *Journal of Membrane Science* 123 (2): 151–56. https://doi.org/10.1016/S0376-7388(96)00180-9.

Hollingsworth Scott A., and Ron O. Dror. 2018. "Molecular Dynamics Simulation for All." *Neuron* 99(6) (1): 1129–43. https://doi.org/10.1016/j.neuron.2018.08.011. Molecular.

Hughes, Zak E., and Julian D. Gale. 2010. "A Computational Investigation of the Properties of a Reverse Osmosis Membrane." *Journal of Materials Chemistry* 20 (36): 7788–99.

Hughes, Zak E., Louise A. Carrington, Paolo Raiteri, and Julian D. Gale. 2011. "A Computational Investigation into the Suitability of Purely Siliceous Zeolites as Reverse Osmosis Membranes." *The Journal of Physical Chemistry C* 115 (10): 4063–75.

Hwang, Sun-Tak. 2004. "Nonequilibrium Thermodynamics of Membrane Transport." *AIChE Journal* 50 (4): 862–70.

Ivetac, Anthony, and Mark S. P. Sansom. 2008. "Molecular Dynamics Simulations and Membrane Protein Structure Quality." *European Biophysics Journal* 37 (4): 403–9.

Jatkar, Krishnadeo, Jae W. Lee, and Sangyong Lee. 2010. "Determination of Reference Chemical Potential Using Molecular Dynamics Simulations." *Journal of Thermodynamics* 2010 (Md): 1–5. https://doi.org/10.1155/2010/342792.

Jia, Yu-xiang, Hai-lan Li, Meng Wang, Lian-ying Wu, and Yang-dong Hu. 2010. "Carbon Nanotube: Possible Candidate for Forward Osmosis." *Separation and Purification Technology* 75 (1): 55–60. https://doi.org/10.1016/j.seppur.2010.07.011.

Jiang, Chi, Lei Tian, Yingfei Hou, and Q. Jason Niu. 2019. "Nanofiltration Membranes with Enhanced Microporosity and Inner-Pore Interconnectivity for Water Treatment: Excellent Balance between Permeability and Selectivity." *Journal of Membrane Science* 586 (March): 192–201. https://doi.org/10.1016/j.memsci.2019.05.075.

Jong, Djurre H. De, Lars V. Schäfer, Alex H. De Vries, Siewert J. Marrink, Herman J. C. Berendsen, and Helmut Grubmüller. 2011. "Determining Equilibrium Constants for Dimerization Reactions from Molecular Dynamics Simulations." *Journal of Computational Chemistry* 32 (9): 1919–28.

Jung, Da Hee, Jijung Lee, Do Yeon Kim, Young Geun Lee, Minkyu Park, Sangho Lee, Dae Ryook Yang, and Joon Ha Kim. 2011. "Simulation of Forward Osmosis Membrane

Process: Effect of Membrane Orientation and Flow Direction of Feed and Draw Solutions." *Desalination* 277 (1–3): 83–91. https://doi.org/10.1016/j.desal.2011.04.001.

Kalra, Amrit, Shekhar Garde, and Gerhard Hummer. 2003. "Osmotic Water Transport through Carbon Nanotube Membranes." *Proceedings of the National Academy of Sciences* 100 (18): 10175–80.

Khalili-araghi, Fatemeh, James Gumbart, Po-chao Wen, Marcos Sotomayor, and Klaus Schulten. 2010. "Molecular Dynamics Simulations of Membrane Channels and Transporters." *Curr Opin Struct Biol* 19 (2): 128–37. https://doi.org/10.1016/j.sbi.2009.02.011. Molecular.

Kim, Daejoong, and Eric Darve. 2009. "High-Ionic-Strength Electroosmotic Flows in Uncharged Hydrophobic Nanochannels." *Journal of Colloid and Interface Science* 330 (1): 194–200.

Kim, Jihye, Jijung Lee, and Joon Ha Kim. 2012. "Overview of Pressure-Retarded Osmosis (PRO) Process and Hybrid Application to Sea Water Reverse Osmosis Process." *Desalination and Water Treatment* 43 (1–3): 193–200.

Kim, Young M., Seung J. Kim, Yong S. Kim, Sangho Lee, In S. Kim, and Joon Ha Kim. 2009. "Overview of Systems Engineering Approaches for a Large-Scale Seawater Desalination Plant with a Reverse Osmosis Network." *Desalination* 238 (1–3): 312–32.

Kopeć, Wojciech, Jelena Telenius, and Himanshu Khandelia. 2013. "Molecular Dynamics Simulations of the Interactions of Medicinal Plant Extracts and Drugs with Lipid Bilayer Membranes." *FEBS Journal* 280 (12): 2785–2805. https://doi.org/10.1111/febs.12286.

Kotelyanskii, M. J., N. J. Wagner, and M. E. Paulaitis. 1998. "Atomistic Simulation of Water and Salt Transport in the Reverse Osmosis Membrane FT-30." *Journal of Membrane Science* 139 (1): 1–16.

Kumar, Hemant, and Prabal K. Maiti. 2011. "Introduction to Molecular Dynamics Simulation." In, 23:161–97. https://doi.org/10.1007/978-93-86279-50-7_6.

Kumar, Pratik, Satinder Kaur Brar, and Maximiliano Cledon. 2022. "A Computational Fluid Dynamics Approach to Predict the Scale-up Dimension of a Water Filter Column." *Case Studies in Chemical and Environmental Engineering* 5 (March): 100201. https://doi.org/10.1016/j.cscee.2022.100201.

Luo, Yun, and Benoît Roux. 2010. "Simulation of Osmotic Pressure in Concentrated Aqueous Salt Solutions." *The Journal of Physical Chemistry Letters* 1 (1): 183–89.

Luo, Yun, Edward Harder, Ron S. Faibish, and Benoît Roux. 2011a. "Computer Simulations of Water Flux and Salt Permeability of the Reverse Osmosis FT-30 Aromatic Polyamide Membrane." *Journal of Membrane Science* 384 (1–2): 1–9. https://doi.org/10.1016/j.memsci.2011.08.057.

Luo, Yun, Edward Harder, Ron S. Faibish, and Benoît Roux. 2011b. "Computer Simulations of Water Flux and Salt Permeability of the Reverse Osmosis FT-30 Aromatic Polyamide Membrane." *Journal of Membrane Science* 384 (1–2): 1–9.

Mallada, Reyes, and Miguel Menéndez. 2008. *Inorganic Membranes: Synthesis, Characterization and Applications.* Elsevier.

Mao, Yijin, and Yuwen Zhang. 2012. "Thermal Conductivity, Shear Viscosity and Specific Heat of Rigid Water Models." *Chemical Physics Letters* 542 (July): 37–41. https://doi.org/10.1016/j.cplett.2012.05.044.

Membrane and Desalination Technologies. 2008. *Membrane and Desalination Technologies.* https://doi.org/10.1007/978-1-59745-278-6.

Mi, Baoxia, David G. Cahill, and Benito J. Mariñas. 2007. "Physico-Chemical Integrity of Nanofiltration/Reverse Osmosis Membranes during Characterization by Rutherford Backscattering Spectrometry." *Journal of Membrane Science* 291 (1–2): 77–85. https://doi.org/10.1016/j.memsci.2006.12.052.

Mohammadi, Toraj, and Ashkan Esmaeelifar. 2004. "Wastewater Treatment Using Ultrafiltration at a Vegetable Oil Factory." *Desalination* 166 (1–3): 329–37. https://doi.org/10.1016/j.desal.2004.06.087.

Moonkhum, Monruedee, Young Geun Lee, Yun Seok Lee, and Joon Ha Kim. 2010. "Review of Seawater Natural Organic Matter Fouling and Reverse Osmosis Transport Modeling for Seawater Reverse Osmosis Desalination." *Desalination and Water Treatment* 15 (1–3): 92–107.

Mótyán, János, Ferenc Tóth, and József Tőzsér. 2013. "Research Applications of Proteolytic Enzymes in Molecular Biology." *Biomolecules* 3 (4): 923–42. https://doi.org/10.3390/biom3040923.

Murad, S. 1996. "Molecular Dynamics Simulations of Osmosis and Reverse Osmosis in Solutions." *Adsorption* 2 (1): 95–101.

Murad, Sohail. 2011. "The Role of External Electric Fields in Enhancing Ion Mobility, Drift Velocity, and Drift–Diffusion Rates in Aqueous Electrolyte Solutions." *The Journal of Chemical Physics* 134 (11): 114504.

Murad, Sohail, and Ludwig C. Nitsche. 2004. "The Effect of Thickness, Pore Size and Structure of a Nanomembrane on the Flux and Selectivity in Reverse Osmosis Separations: A Molecular Dynamics Study." *Chemical Physics Letters* 397 (1–3): 211–15. https://doi.org/10.1016/j.cplett.2004.08.106.

Park, Minkyu, Ji Jung Lee, Sangho Lee, and Joon Ha Kim. 2011. "Determination of a Constant Membrane Structure Parameter in Forward Osmosis Processes." *Journal of Membrane Science* 375 (1–2): 241–48. https://doi.org/10.1016/j.memsci.2011.03.052.

Parvatiyar, Madan G. 1998. "Entropy Generation in Ultrafiltration Processes." *Journal of Membrane Science* 144 (1–2): 125–32. https://doi.org/10.1016/S0376-7388(98)00041-6.

Pascal, Tod A., William A. Goddard, and Yousung Jung. 2011. "Entropy and the Driving Force for the Filling of Carbon Nanotubes with Water." *Proceedings of the National Academy of Sciences* 108 (29): 11794–98.

Pitman, Michael C., Alan Grossfield, Frank Suits, and Scott E. Feller. 2005. "Role of Cholesterol and Polyunsaturated Chains in Lipid– Protein Interactions: Molecular Dynamics Simulation of Rhodopsin in a Realistic Membrane Environment." *Journal of the American Chemical Society* 127 (13): 4576–77.

Pourmousa, Mohsen, and Richard W. Pastor. 2018. "Molecular Dynamics Simulations of Lipid Nanodiscs." *Biochimica et Biophysica Acta–Biomembranes* 1860 (10): 2094–2107. https://doi.org/10.1016/j.bbamem.2018.04.015.

Riviere, John C., and Sverre Myhra. 2009. *Handbook of Surface and Interface Analysis: Methods for Problem-Solving*. CRC press.

Róg, Tomasz, Mykhailo Girych, and Alex Bunker. 2021. "Mechanistic Understanding from Molecular Dynamics in Pharmaceutical Research 2: Lipid Membrane in Drug Design." *Pharmaceuticals* 14 (10). https://doi.org/10.3390/ph14101062.

Samaei, Seyed Mohsen, Shirley Gato-Trinidad, and Ali Altaee. 2018. "The Application of Pressure-Driven Ceramic Membrane Technology for the Treatment of Industrial Wastewaters–A Review." *Separation and Purification Technology* 200 (October 2017): 198–220. https://doi.org/10.1016/j.seppur.2018.02.041.

Saurabh, Suman, Ponnurengam Malliappan Sivakumar, Venkatesan Perumal, Arezoo Khosravi, Abimanyu Sugumaran, and Veluchamy Prabhawathi. 2020. "Molecular Dynamics Simulations in Drug Discovery and Drug Delivery." *Engineering Materials*, 275–301. https://doi.org/10.1007/978-3-030-36260-7_10.

Schlick, Tamar. 2010. *Molecular Modeling and Simulation: An Interdisciplinary Guide*. Vol. 2. Springer.

Seyyedattar, Masoud, Sohrab Zendehboudi, and Stephen Butt. 2019. "Molecular Dynamics Simulations in Reservoir Analysis of Offshore Petroleum Reserves: A Systematic Review of Theory and Applications." *Earth-Science Reviews* 192 (February): 194–213. https://doi.org/10.1016/j.earscirev.2019.02.019.

Sharma, Bhanuday, and Rakesh Kumar. 2019. "Estimation of Bulk Viscosity of Dilute Gases Using a Nonequilibrium Molecular Dynamics Approach." *Physical Review E* 100 (1): 013309. https://doi.org/10.1103/PhysRevE.100.013309.

She, F. H., W. M. Gao, Z. Peng, P. D. Hodgson, and L. X. Kong. 2008. "Micro–and Nano-Characterization of Membrane Materials." *Journal of the Chinese Institute of Chemical Engineers* 39 (4): 313–20. https://doi.org/10.1016/j.jcice.2008.01.008.

Šostar-Turk S., M. Simonič, and I. Petrinić. 2005. "Wastewater Treatment after Reactive Printing." *Dyes and Pigments* 64 (2): 147–52. https://doi.org/10.1016/j.dyepig.2004.04.001.

Srivastava, Ashutosh, Tetsuro Nagai, Arpita Srivastava, Osamu Miyashita, and Florence Tama. 2018. "Role of Computational Methods in Going beyond X-Ray Crystallography to Explore Protein Structure and Dynamics." *International Journal of Molecular Sciences* 19 (11). https://doi.org/10.3390/ijms19113401.

Starr, Francis W., Stephen Harrington, Francesco Sciortino, and H. Eugene Stanley. 1999. "Dynamics of Simulated Water under Pressure." *Physical Review Letters* 82 (18): 3629–32. https://doi.org/10.1103/PhysRevLett.82.3629.

Suk, M. E., A. V. Raghunathan, and N. R. Aluru. 2008. "Fast Reverse Osmosis Using Boron Nitride and Carbon Nanotubes." *Applied Physics Letters* 92 (13): 133120.

Sun, Lequn, Weichen Lin, Xiaotian Wu, Johny Cabrera, Daoyi Chen, and Xia Huang. 2022. "Deciphering the Spatial Fouling Characteristics of Reverse Osmosis Membranes for Coal Chemical Wastewater Treatment." *Separation and Purification Technology* 286 (January): 120456. https://doi.org/10.1016/j.seppur.2022.120456.

Suwaileh, Wafa Ali, Daniel James Johnson, Sarper Sarp, and Nidal Hilal. 2018. "Advances in Forward Osmosis Membranes: Altering the Sub-Layer Structure via Recent Fabrication and Chemical Modification Approaches." *Desalination.* Elsevier. https://doi.org/10.1016/j.desal.2018.01.035.

Takht Ravanchi, Maryam, Tahereh Kaghazchi, and Ali Kargari. 2009. "Application of Membrane Separation Processes in Petrochemical Industry: A Review." *Desalination* 235 (1–3): 199–244. https://doi.org/10.1016/j.desal.2007.10.042.

Thompson, Aidan P., and Grant S. Heffelfinger. 1999. "Direct Molecular Simulation of Gradient-Driven Diffusion of Large Molecules Using Constant Pressure." *The Journal of Chemical Physics* 110 (22): 10693–705. https://doi.org/10.1063/1.478996.

Uchida, Hirohisa, and Masakuni Matsuoka. 2004. "Molecular Dynamics Simulation of Solution Structure and Dynamics of Aqueous Sodium Chloride Solutions from Dilute to Supersaturated Concentration." *Fluid Phase Equilibria* 219 (1): 49–54. https://doi.org/10.1016/j.fluid.2004.01.013.

Wang, Luying, Randall S. Dumont, and James M. Dickson. 2012. "Nonequilibrium Molecular Dynamics Simulation of Water Transport through Carbon Nanotube Membranes at Low Pressure." *The Journal of Chemical Physics* 137 (4): 44102.

Wang, Yanhui, Xu Chen, Jinchang Zhang, Jingmei Yin, and Huanmei Wang. 2009. "Investigation of Microfiltration for Treatment of Emulsified Oily Wastewater from the Processing of Petroleum Products." *Desalination* 249 (3): 1223–27. https://doi.org/10.1016/j.desal.2009.06.033.

Yeap, Swee Pin, JitKang Lim, Boon Seng Ooi, and Abdul Latif Ahmad. 2017. "Agglomeration, Colloidal Stability, and Magnetic Separation of Magnetic Nanoparticles: Collective

Influences on Environmental Engineering Applications." *Journal of Nanoparticle Research* 19 (11): 368. https://doi.org/10.1007/s11051-017-4065-6.

Yordanov, D. 2010. "Preliminary Study of the Efficiency of Ultrafiltration Treatment of Poultry Slaughterhouse Wastewater." *Bulgarian Journal of Agricultural Science* 16 (6): 700–704.

Yoshioka, Tomohisa, Keisuke Kotaka, Keizo Nakagawa, Takuji Shintani, Hao-Chen Wu, Hideto Matsuyama, Yu Fujimura, and Takahiro Kawakatsu. 2018. "Molecular Dynamics Simulation Study of Polyamide Membrane Structures and RO/FO Water Permeation Properties." *Membranes* 8 (4): 127. https://doi.org/10.3390/membranes8040127.

Zhu, Fangqiang, and Klaus Schulten. 2003. "Water and Proton Conduction through Carbon Nanotubes as Models for Biological Channels." *Biophysical Journal* 85 (1): 236–44. https://doi.org/10.1016/S0006-3495(03)74469-5.

Zhu, Fangqiang, Emad Tajkhorshid, and Klaus Schulten. 2001. "Molecular Dynamics Study of Aquaporin-1 Water Channel in a Lipid Bilayer." *FEBS Letters* 504 (3): 212–18. https://doi.org/10.1016/S0014-5793(01)02749-1. 2004. "Theory and Simulation of Water Permeation in Aquaporin-1." *Biophysical Journal* 86 (1 I): 50–57. https://doi.org/10.1016/S0006-3495(04)74082-5.

3 Membrane Process for the Water Purification CFD Approach

Ravikant R. Gupta[1], Vineet Kumar[2], and Richa Agarwal[1]
[1]Department of Chemical Engineering, Banasthali Vidyapith, Rajasthan, India
[2]Department of Chemical Engineering, IIT-ISM Dhanbad, Jharkhand, India

CONTENTS

3.1 Introduction 40
3.2 Membrane Separation Processes 40
    3.2.1 Reverse Osmosis (RO) 40
    3.2.2 Nanofiltration (NF)..41
    3.2.3 Microfiltration (MF)..41
    3.2.4 Ultrafiltration (UF) 42
    3.2.5 Dialysis..42
3.3 Membrane Modules...42
    3.3.1 Tubular Module 43
    3.3.2 Spiral-Wound Module (SWM)..43
    3.3.3 Hollow-Fiber Module (HF) 44
3.4 Types of Membrane...45
3.5 Governing Equation...46
3.6 Mathematical Model..47
    3.6.1 Pore Flow model 47
    3.6.2 Film Theory Model 48
    3.6.3 Spiegler–Kedem Model 48
    3.6.4 Gel Polarization Model 49
    3.6.5 Osmotic Pressure Model 49
    3.6.6 Resistance in Series Model (RIS)..50
3.7 Computational Fluid Dynamics (CFD) in Membrane Process 50
3.8 CFD Modeling and Mass Transfer in Gel Layer............................52
3.9 CFD-Based Membrane Processes in Wastewater Treatment.....................55
3.10 Conclusions 55
References..56

3.1 INTRODUCTION

A rapid increase in population and industrialization increased the demand for clean and safer water. The source of drinkable water is only 3%, and about 97% of water is in oceans, which cannot be used directly due to high salt content. Only a few countries like UAE and Qatar, with limited freshwater sources, utilize the desalination and water purification systems extensively to provide freshwater (Yang et al., 2012). For the rest of the world, the major issue is the contamination of available water by waste generated from industrial, municipal, and agricultural activity. Contaminated water requires pretreatment processes like adsorption, filtration, and distillation before its application.

Nearly 50 years of search for a feasible alternative to traditional separation methods, such as absorption and distillation, has led to synthetic membrane-based separation methods. Membrane-based technology is generally clean and requires less capital than conventional techniques in many areas. The growing market of membranes of various types has drawn interest in conducting more research and study to develop efficient design strategies that aim to enhance the overall performance of membrane separation techniques (Kumar et al., 1998). Currently, membrane processes like microfiltration (MF), ultrafiltration (UF), nanofiltration (NF), and reverse osmosis (RO) are used to treat various sources of water. The membrane process can be used individually or in combination with two or more processes (conventional, non-conventional, or hybrid) for better and more effective wastewater treatment. One of the most successful integrated systems is membrane bioreactor (MBR), which combines membrane filtration with bacterial treatment.

3.2 MEMBRANE SEPARATION PROCESSES

In most membrane separation techniques, the feed stream is split into two output streams, i.e., permeate (material that passes through the membrane) and the retentate (material that does not pass through the membrane, the reject). Figure 3.1 shows various membrane processes based on particle size and the primary factors affecting the process of separation. The processes (RO, NF, UF, MF, dialysis, and electrodialysis) cover a wide range of particle sizes and are capable of removing both ions and macromolecules from solutions (Kumar et al., 2012).

3.2.1 REVERSE OSMOSIS (RO)

Reverse osmosis (RO) is becoming one of the popular methods for water treatment. It is a pressure-driven process having a pore size of around 0.0001 microns. The term RO is applied to water purification from an aqueous solution by passing through a membrane that ideally should only be permeable to water (Santoyo et al., 2004). In this process, the pressure higher than the osmotic pressure of solute is applied (increasing the solvent's chemical potential), forcing the solvent molecules to permeate through the membrane pores. Applications of the RO process include removal of inorganic and organic pollutants from wastewater, purification of products in the

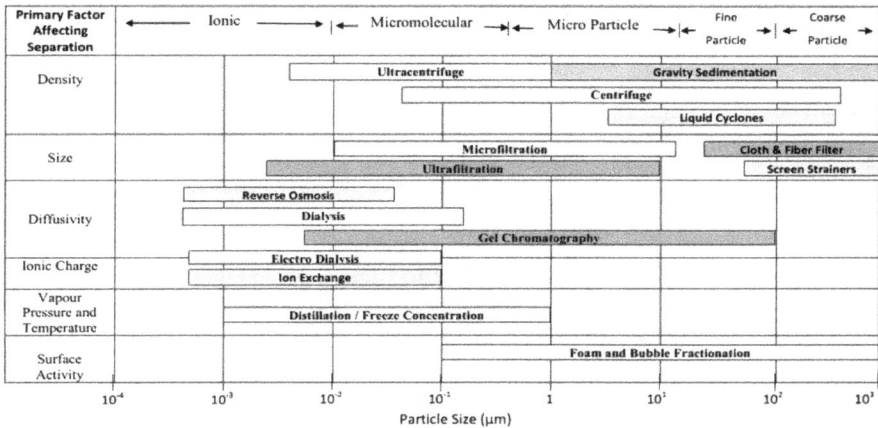

FIGURE 3.1   Types of membrane separation process based on particle size ranges.

chemical, pharmaceutical, desalination of brackish water from the sea, food industries, etc.

3.2.2 NANOFILTRATION (NF)

Properties of nanofiltration lie between those of RO and UF. The NF technology is needed because the RO membranes completely retain all the dissolved salt ions, and UF rejects any solute higher than 10000 gm-mole. So to retain solute in the range of 500 to 10000 gm-moles (e.g., calcium and magnesium), NF technology is needed. The pore size of the NF membrane is around 0.001 micron, with a driving force of 100–250 kPa pressure (Shon et al., 2013). The demand for NF has increased exponentially in drinking water and wastewater treatment as it removes the calcium and magnesium ions resulting in a decrease in the hardness of water (Abdel-Fatah, 2018). The NF works by both convective and solution-diffusion mechanisms. The advantage of nanofiltration is that it removes most of the organic molecules, viruses, natural organic matter, and a range of salts (Childress and Elimelech, 2000). The limitation of NF is that it consumes high energy, and the membrane requires frequent replacement due to dissolved solids.

3.2.3 MICROFILTRATION (MF)

Microfiltration processes are used in many industrial applications where particles of a size range from about 0.1 to 10 microns must be retained. Microfiltration membranes consist of the largest membrane pores compared to the NF and RO. The osmotic pressure for MF is negligible, and the required driving force, i.e., trans-membrane pressure (TMP), is relatively small. MF is commonly used as prefiltration in water treatment for turbidity reduction by removing suspended solids (Mulder, 1991). MF fills the gap between conventional filtration and ultrafiltration.

3.2.4 ULTRAFILTRATION (UF)

UF is mainly used to separate solute particles like macromolecular solutes and colloidal suspensions from wastewater in a size range of about 0.001–0.1 microns (Filipe and Ghosh, 2005; Bakhshayeshi et al., 2012). The membrane used in UF is less dense than those used in RO. Therefore, high permeate fluxes can be achieved using much smaller TMP (100–500 kPa) (Bhattacharjee and Bhattacharya, 1992). UF is used as a pretreatment process for secondary sewage treatment to reduce turbidity and particulate matter in the feed water stream. The membrane used for UF has an asymmetric structure with a denser skin layer leading to higher hydrodynamic resistance. The separation process in UF depends upon pore diameter, membrane-solute interaction, and concertation polarization (Singh and Hankis, 2016)

3.2.5 DIALYSIS

Dialysis is perhaps the most successful and widely used membrane separation process ranging from hemodialysis (Cancilla et al., 2022) to wastewater treatment (Kavitha et al., 2022) used to separate colloids from dissolved molecules or ions, and acids from its salt solutions (diffusion dialysis). In dialysis, macro solutes are circulated on one side of the membrane, and the solute separation is driven by a difference in concentration between the two sides of the membrane (forward osmosis). Dialysis is a slow process, and its rate can be altered by heating or by applying an electric field called electrodialysis. In the electrodialysis, electrodes are placed with the membrane, and due to electric potential difference, ions travel across the membrane. In wastewater treatment, it is mostly used to purify brackish water, and RO reject. The dialysis process is best suitable for a feed stream having TDS <500.

3.3 MEMBRANE MODULES

The success of a membrane process depends on the design of the membrane module. A number of different membrane modules have been designed (using flat sheet and tube-shaped membranes) to achieve a high degree of separation, keeping the two most important design criteria in mind. Firstly, better contact between the membrane and the fluid flowing through the module for efficient mass transfer. Secondly, the module must provide a large active membrane area available for separation per unit volume so that the plant size and capital cost are minimum.

Other important criteria include hydrodynamics of the module, high mechanical stability, ease of maintenance, life span of the membrane. The available modules in the process industries are plate-and-frame modules, tubular modules, hollow-fiber, and spiral-wound modules. Comparisons of different membrane modules are summarized in Table 3.1. Plate and frame modules have the least packing density (surface area per unit volume) but can be fabricated easily with widely different characteristic parameters. However, hollow fiber and spiral wound membrane replaces tubular module and plate and frame, respectively, due to their relatively high cost and less efficiency.

TABLE 3.1
Comparison Between Modular Configurations (Mulder, 1991; Baker, 2004)

Parameter	Module Tubular	Spiral-wound	Hollow Fiber
Specific surface area (m^2/m^3)	300	1000	15000
Inner diameter or spread (m)	0.02–0.05	0.005–0.02	0.0005–0.002
Flux (L/m^2 day)	300–1000	300–1000	30–100
Production (m^3/m^3 per module per/day)	100–1000	300–1000	450–1500
Space velocity (m/s)	1–5	0.25–0.5	0.005
Pressure loss (kPa)	200–300	100–200	30
Pretreatment	Simple	Medium	High
Plugging	Small	Medium	Elevated
Replacement	Easy	Difficult	Impossible
Cleaning: Mechanical	Possible	Not possible	Not possible
Chemical	Possible	Possible	Possible

3.3.1 TUBULAR MODULE

The tubular membrane is located inside porous (or perforated) support material, which is placed inside a pressure-tight tube shell (12 to 25 mm in dia). The feed stream usually flows within the tube while the product permeates comes radially out through the membrane surface. Tubular modules are easy to maintain but their main disadvantage is the small mass transfer area as compared to the spiral and hollow fiber membranes (Baker, 2004). The tubular membrane is well suited for wastewater with high suspended solids as these modules are easy to clean. The tubular module has the advantage of less fouling and is easy to clean with harsh chemicals, backwash, and even mechanical cleaning, which might not be suitable for the other configurations. The disadvantages are low packing density and large size in comparison to other modules.

3.3.2 SPIRAL-WOUND MODULE (SWM)

The basic concept for designing this module is to bundle a large surface area into a small volume. Spiral-wound modules (SWM) are essentially flat membrane sheets placed between highly porous perforated supports, and then the sheet is rolled over a perforated tube, forming a spiral and placed in a pressure-tight enclosure (Senthilmurugan et al., 2005). The schematic configuration of the spiral wound membrane is shown in Figure 3.2(a-b). The aqueous feed solution is passed through the pipe so that it flows into the perforated mesh screen along the membrane surface. Permeates flow into the interior of the membrane envelopes and flows along the spiral path towards the central tube. However, because of the thin size of the flow channel, the chances of plugging are high (Kumar et al., 2004). The SWM is the standard module type used in the RO/NF industry. SWMs have a very high packing density

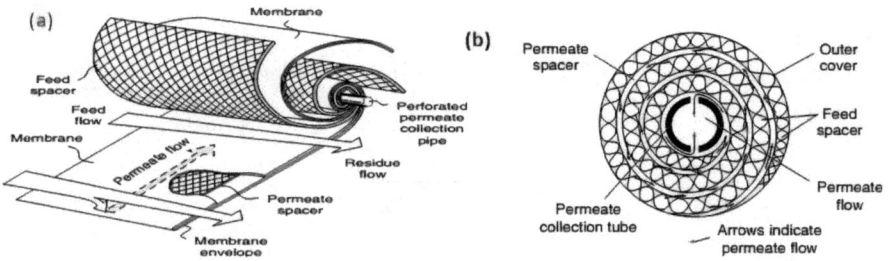

FIGURE 3.2   General design features of a spiral-wound membrane module. (a) Spiral-wound module. (b) Cross-sectional view of spiral wound module (Baker, 2002).

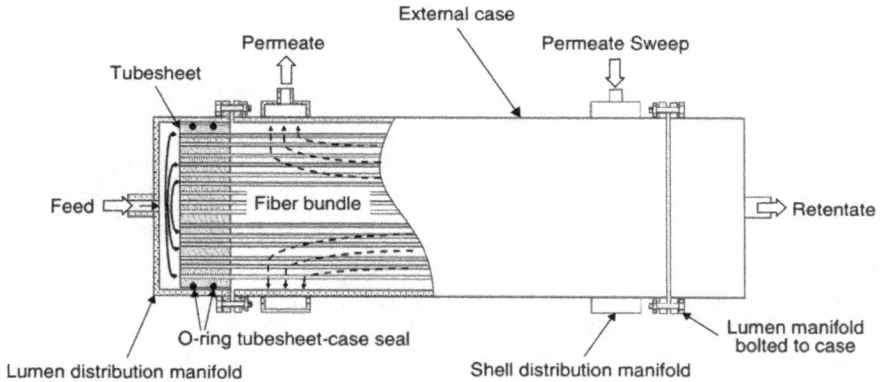

FIGURE 3.3   A hollow fiber membrane module lumen feed and shell sweep operating in counter-current contacting mode (Mat et al., 2014).

in comparison to plate and frame, and tubular configurations. But fouling tendency is greater than the tubular and plate and frame module. It allows for easy in-situ cleaning through backwash methods.

3.3.3 HOLLOW-FIBER MODULE (HF)

Hollow-fiber modules contain hundreds of fine membrane fiber tubes bundled together in either a straight configuration or in a U-shape configuration housed in a pressure vessel (Chatterjee et al., 2004). The hollow fibers' diameter varies over a wide range, from 50 to 3000 μm (Baker, 2004). The HF membranes are mostly of microporous structure with a dense selective layer either on the outside or inside the fiber (Belfort, 1988). A cross-flow hollow fiber membrane module configuration is shown in Figure 3.3. The feed configuration depends upon the diameter of the fiber. For small diameter fibers, the feed is introduced through outside, and permeate is collected down the fiber bore (lumen). For larger diameter (200–500 μm), the feed is introduced to the lumen, and permeate is collected from the outer shell. The HF membranes are now used in seawater desalination plants and industrial effluent treatment plants. Hollow fiber features a very large packing density because of the small diameter of the fiber tube. The high flexibility of the tubes is an additional advantage to certain filter configurations that cannot be achieved in other filtration configurations (Ghosh et al., 2000a). Fouling and easy breakage of tubes are the main disadvantages of HFM.

3.4 TYPES OF MEMBRANE

A porous membrane layer is defined as a thick or thin layer of material of pore size 1–100 nm, which act as a barrier to the flow of certain molecular species larger in size than the membrane pores present in the liquid when a suitable driving force (concentration, temperature or pressure gradient) is applied across the membrane (Miroslav et al., 2003; Goksen, 2005). Based on the above definition, a large number of materials and structures may be considered as a membrane. However, membrane materials are generally organic (e.g., polyethylene, polysulfone polypropylene, and cellulose acetate) or inorganic (ceramics, metals, or silica). Most industrial membranes are currently based on polymeric materials though the inorganic membrane is gaining attention and may prove to be the future membrane material. The industrial application of inorganic materials is increasing nowadays due to their chemical, mechanical, and thermal robustness and reusability. The ceramic membranes along with photocatalytic material are also recently gaining considerable interest in wastewater and groundwater treatment process. The basic requirement of a membrane for its industrial applications is high flux, high perm-selectivity, high mechanical strength, and ease of fabrication.

Separation phenomena by a membrane surface layer largely depend upon the following:

(a) Pore size and pore size distribution: influences permeate flux and membrane fouling.
(b) Hydrophilic-hydrophobic nature: it influences interactions between membrane and solute, fouling behavior, and cleaning mechanism.
(c) Surface charge: influences fouling and separation.
(d) Chemical compatibility: affects cleaning and separation operation.
(e) Ease of fabrication and total cost: influences economics and viability.

On the basis of structure and cross-section, membranes are classified into two types: isotropic and anisotropic. The homogenous composition such as porous, dense film non-porous and electrically charged membranes are isotropic type. In porous membrane solute separation is based on particulate size and pore dia. In the case of non-porous separation is due to diffusivity with respect to driving force. The anisotropic membranes are classified as composite membranes (coated film and thin film). Anisotropic membranes are generally used in RO, NF, and UF. Table 3.2 shows various materials used for the preparation of membranes.

TABLE 3.2
Membrane Processes, Membrane Preparation, and Material Used for Preparation

Process	Membrane Preparation	Membrane Material	Membrane Type
RO/NF	• Phase inversion process	TFC, cellulose acetate (CA), polyamide (PA)	Semi porous, asymmetric, thin film composite
UF	• Polymer membrane: phase inversion process • Ceramic membranes: sintering and sole gel processes.	Polysulphone (PS), polyvinylpyrrolidone (PVP) polyether sulphone (PES), polyacrylonitrile (PAN), and polyvinylidene fluoride (PVDF), CA,	Micro porous, asymmetric
MF	• Polymer membrane by sintering, track-etching, stretching, or by phase inversion. • Ceramic membranes: sintering and sole gel processes.	PTFE (Teflon), PVDF, PS, cellulosic, polyester, polypropylene (PP), polycarbonate, polyether imide and nylon 6.	Porous, asymmetric or symmetric

3.5 GOVERNING EQUATION

To explain the behavior of a membrane separation process system, it is helpful to first define relevant parameters. The first and foremost of these is the permeate flux, J_v (m³/m².s), also called permeation velocity, V_w (m/s).

$$J_v = \frac{dV}{A_{sm} \cdot dt}$$ (3.1)

where A_{sm} is the membrane surface area, dV is the infinitesimal amount of permeate volume collected during the infinitesimal time interval dt.

The rejection coefficient of a membrane R_i is defined as the efficiency with which the membrane can separate the solute in the feed mixture.

$$R_i = 1 - \frac{C_p}{C_b}$$ (3.2)

where, C_p is the permeate stream concentration, and C_b is the bulk feed concentration. The value of R_i varies between zero (free to pass through membrane layer) and one (complete solute retention by membrane surface).

The flow through the membrane is characterized either by the hydrodynamic permeability L_p (m/Pa.s), which is defined as

$$L_p = \frac{J_v}{\Delta P} \qquad (3.3a)$$

or by membrane permeability (also called intrinsic permeability) κ (m), the reciprocal of membrane resistance R, given by

$$\kappa = \frac{\mu.J_v}{\Delta P} \qquad (3.3b)$$

In the above equations, ΔP is the transmembrane pressure difference (TMP), i.e., feed side pressure minus permeate side pressure, J_v is the permeate flux, and μ is the viscosity.

3.6 MATHEMATICAL MODEL

A number of models are available to study the phenomena of flux and fouling on membrane processes. A few of them, being widely used for membrane flux and fouling prediction, are presented below.

3.6.1 PORE FLOW MODEL

For membranes with uniformly distributed pores, to estimate the pure solvent permeate flux (i.e., permeation with no fouling and negligible concentration polarization), generally following Hagen–Poiseuille equation is used.

$$V_w = \frac{\pi \varepsilon d_p^4 \Delta P}{32 \delta_M \mu} \qquad (3.4)$$

where, μ is the viscosity of permeating fluid, ε_p is the membrane porosity, δ_M is the membrane thickness, and d_p is the pore diameter. The above Hagen–Poiseuille equation assumes that fluid density is constant, steady-state condition prevails, flow is laminar, and fluid is a Newtonian fluid. The equation is further simplified by clubbing all the constant parameters (i.e., $32\delta M / \pi \varepsilon d_p^4$) as the membrane resistance R giving,

$$V_w = \Delta P / (R\mu) \qquad (3.5)$$

Eqn. 3.5, and its many other variants, are widely used in calculating permeate flux.

3.6.2   Film Theory Model

The film theory model is one of the simplest theories for modeling flux. The concentration profile is described by a 1D mass balance equation with the assumption that the axial convection of solute near the membrane surface is negligible. The convective flux of solute normal to the membrane surface is balanced by the back diffusion. This model is valid only in pressure-independent regions (or at a very low TMP) since there is no pressure-dependent term in the film theory model.

$$V_w = (k)\ln\frac{C_w - C_p}{C_b - C_p} \tag{3.6}$$

$$\text{where } k = \frac{D}{\delta} \tag{3.7}$$

k is the mass transfer coefficient, D and δ are the molecular diffusivity of the solute in the solvent and effective film thickness, C_b, C_p, and C_w are the solute concentration, in the bulk phase, in the permeate stream and at the the the membrane surface, respectively.

3.6.3   Spiegler–Kedem Model

The S–K model (Spiegler and Kedem, 1966) is based on the concept of irreversible thermodynamics, which is used to derive transport equations for solute membrane interaction in the NF and RO processes. The membrane performance is calculated using two coefficients: reflection coefficient (σ) and solute permeability (P_s). The coefficient's values are obtained from the experimental data of rejection versus permeation flux values. Later they are fit to the analytical solution of the Spiegler–Kedem model (Ahmed, 2013). The equations for the nonlinear Spiegler–Kedem model for intrinsic rejection (R) and permeate flux, V_w and are as follows (Sarkar et al., 2006):

$$V_w = L_p\left(\Delta p - \sigma\Delta\pi\right) \tag{3.8}$$

$$R = 1 - \frac{C_p}{C_m} = \frac{\sigma(1-F)}{1-\sigma F} \tag{3.9}$$

where F is the flow parameter defined as below:

$$F = \exp\left(-\frac{(1-\sigma)V_w}{P_s}\right) \tag{3.10}$$

The basic difference between the Spiegler–Kedem model and the film theory model is that this model incorporates the reflection coefficient (σ) with a value ranging from

0 to 1. The reflection coefficient represents the rejection capability of a membrane. For the case where $\sigma = 1$, Spiegler–Kedem model assumes the membrane as an ideal RO membrane where no convective transport takes place. On the other hand, $\sigma = 0$ represents an entirely unselective membrane where no rejection is assumed on the membrane surface. For MF and UF membranes, which have pores, the value of σ is a positive, less than one.

3.6.4 GEL POLARIZATION MODEL

The basic assumption of this model is that due to concentration polarization, the concentration at the membrane interface cannot exceed beyond a fixed value (C_G) at which solute behaves like a gelly. With further increase in pressure results in an increase in thickness and compaction of this layer of maximum concentration (known as gel layer) such that there is no increase, rather some decrease in flux with increasing pressure is observed (Park and Barnett, 2001; Hidane et al., 2022). The gel concentration depends on the shape, size, degree of salvation, chemical structure but is independent of the bulk concentration. The concentration polarization model can be described by the use of the film theory model (Eqn. 3.6) as

$$V_w = k \ln \frac{C_G}{C_B} \tag{3.11}$$

3.6.5 OSMOTIC PRESSURE MODEL

This model assumes that permeate flux decreases due to the osmotic pressure exerted by solutes retained at the membrane layer (Sreenivas et al., 2002). This osmotic pressure is generally small compared to the applied pressure, which is the driving force for the solvent to flow through the membrane (Zeman and Zydney, 1996; Cheng et al., 1998b). The applied pressure should be equal or greater than osmotic pressure to ensure permeation flux. The main contribution to the osmotic pressure of a solution often arises due to the polarization of low molecular weight solutes on the membrane surface. The flux is given in the form of the Darcy equation:

$$J_v = \frac{\Delta P - \pi\left(C_w\right)}{R_o} \tag{3.12}$$

where $\pi(C_w)$ is the osmotic pressure of the solute at wall concentration (C_w). For most macromolecules, the osmotic pressure data can be correlated in terms of the virial expressions

$$\pi(C_w) = A_1 C_w + A_2 C_w^2 + A_3 C_w^3 + \dots\dots \tag{3.13}$$

or, an exponential correlation (Parsegian et al., 1995; Cohen and Highsmith, 1997)

$$\ln(\pi) = a + b(C_w)^c \tag{3.14}$$

where C_w is the concentration near the membrane wall and A_1, A_2, A_3,... are first, second, third, ... virial coefficients and a, b, and c are exponential coefficients, respectively.

3.6.6 RESISTANCE IN SERIES MODEL (RIS)

The RIS model considered that the flux decreases due to various resistance offered to the permeate flux, such as intrinsic membrane resistance (R_m), resistance due to concentration polarisation (R_{cp}), and the resistance of the cake layer (R_c) and adsorption layer (R_{ad}). The RIS model can be applied to both non-porous and porous membranes. This model can be described by Darcy's law, in which permeate flux is inversely proportional to various resistance in series and proportional to driving force (Purkait et al., 2004).

$$V_w = \frac{\Delta p}{\mu \sum R_t} \tag{3.15}$$

The above equation is similar to Ohm's law, $I = V/R$, in which an electric circuit (I) is inversely proportional to the total resistance (R) and directly proportional to the potential difference (V) (Chang et al., 2009).

Depending on the nature of membrane-solute interaction, the total resistance can be subdivided into several other significant resistances in series, for example,

$$R_t = R_m + R_c + R_f \tag{3.16}$$

where, R_m is the membrane resistance, R_c is the resistance offered by concentration polarization layer, and R_f includes all resistance other than concentration polarization and membrane resistance. R_f can be substantially greater than the resistance due to the membrane.

3.7 COMPUTATIONAL FLUID DYNAMICS (CFD) IN MEMBRANE PROCESS

The above-discussed models are mass transfer models that ignore the fluid flow pattern near the membrane surface. However, the nature of fluid flow near the membrane surface significantly alters the concentration polarization and cake layer thickness. Computational fluid dynamics (CFD) predicts flow characteristics inside a membrane module. The basic definition of CFD is solving the partial differential equation, which governs the fluid flow with the simplified discretized algebraic equations (Keir and

Jegatheesan, 2014). In the last few decades, CFD-based models in membrane processing have increasingly been used for better understanding of membrane process as experimental surveys provide only a piece of the overall information. In contrast, CFD provides the option to modify parameters, properties of fluids and the geometry of the flow channel in a flexible but defined way (Parvareh et al., 2011).

Investigation into the hydrodynamics of membrane channels largely began with classical papers such as Berman (1953), Yuan and Finkelstein (1956), Kozinski et al. (1970). They were pioneers of the model based on the Navier–Stokes equation for simulation of laminar flow in a rectangular and tubular channel assuming fully developed axial flow. The axial velocity was assumed constant along the channel length because the permeation rate in the radial direction is very low. These hydrodynamic studies of an incompressible fluid flow assume that the maximum axial velocity of the Hagen–Poiseuille flow exists at the center of the porous channel. Many investigators considered the fully developed flow with uniform suction or injected through the permeable channel wall. The characterization of flow in a porous pipe is determined by wall Reynolds number (Re_w). Berman (1953) reduced the equation of motion to an ordinary differential equation for fluid pressure and velocity as a function of the module dimensions. Applying the perturbation technique, he observed pressure drops significantly in a membrane channel than in a non-membrane channel. The velocity profile was more curved near the porous wall than the Hagen–Poiseuille parabolic profile for the non-porous wall.

Compared to non-porous channels, pressure drop in membrane processes is consequential as it affects the transmembrane permeation flux. Due to this, the next generation of approaches for the study of membrane separation evolved by coupling Navier–Stokes and Darcy's equations, as the validity of Darcy's law in porous media flow is well accepted (Gartling et al., 1997). Darcy's law is often used to set up the porous wall (membrane) boundary conditions, whereas the momentum equations and continuity equation define the hydrodynamic condition of the membrane channel. The accomplishment of the two-pronged approach to the coupled hydrodynamics-mass transfer makes the model computationally involved. Different approaches such as finite element (FEM), finite difference (FDM), and finite volume (FVM) methods have been performed to simulate the combined models.

Nassehi (1998) used FEM to simulate the 2D coupling of the Brinkman equation (in place of Darcy's equation), which relates the pressure gradient in a porous tube and Navier–Stokes equation to define tangential viscous laminar flow in the cross-flow membrane module. Although the Brinkmann equation (Eqn. 3.17) is not suitable for membrane separation, as it is limited to highly porous bodies, some researchers have used it in place of the Darcy equation.

$$\Delta p - \mu \nabla^2 v = \frac{\mu}{K} v \qquad (3.17)$$

Nassehi (1998) resolved this issue by developing a Galerkin FEM scheme based on Lagrange elements. Results demonstrated that the scheme was flexible in application, and the continuity equation was satisfied accurately. Damak et al. (2004) simulated a fluid dynamic model of a cross-flow tubular membrane by the method of coupling, assuming constant permeability of the membrane wall and that the viscosity of the inflow and filtrate are identical. The porous boundary conditions near the membrane wall (i.e., at $r = R$ and $0 \leq x \leq L$) were written as

$$v_r = \frac{\kappa}{\mu} \cdot \frac{\Delta P}{e}, \text{ and } v_x = 0 \qquad (3.18)$$

where e is the thickness of the porous wall, ΔP is TMP, and κ and μ are the intrinsic permeability.

An implicit finite difference scheme is used to discretize and solve the governing equation with boundary conditions assuming fully developed flow at the inlet, no-slip condition at the porous wall, and constant wall permeability. The model proposed by Damak et al. (2004) was successfully validated using experimental data. The results also demonstrated the variation of the velocity profile (axial and radial) with and without wall suction for different Reynolds numbers.

3.8 CFD MODELING AND MASS TRANSFER IN GEL LAYER

The modeling of concentration polarization and fluid flow in tubular with permeable layer is not new. Various researchers' groups studied the more complex and reliable modeling using CFD techniques in order to study insight phenomena taking place at the surface of the membrane (Geraldes et al., 2000; de Pinho et al., 2002; Schausberger et al., 2009; Sarkar et al., 2012; Awasthi, and Kumar, 2017; Gupta et al., 2017). Extensive reviews on the use of CFD in membrane separation can be referred to (Fimbres-Weihs and Wiley, 2010; Keir and Jegatheesan, 2014; Toh et al., 2020). CFD study provides a rigorous and detailed analysis of the transient and local conditions for the solute concentration, fouling, and permeate flux with fewer assumptions. For example, estimates of the transient and local permeate flux and fouling behavior were obtained without requiring assumptions on the polarization layer thickness by Macros et al. (2009). To study the effect of Reynolds and Schmidt number on the concentration polarization layer along the membrane length Pak et al. (2008) simulated a tubular porous cross-flow membrane using staggered grid finite difference technique. The work suggested that membrane performance improves with an increase in axial velocity, which causes an increase in Schmidt number and Reynolds number by decreasing the thickness of the concentration polarization boundary layer. Bhattacharyya et al. (1990) used the finite element method with the Galerkin technique to solve the convection-diffusion equation to determine the concentration profile within the membrane module. A wide range of Reynolds numbers in multi-component solutions was examined for a spiral wound module approximated as a straight channel with a spacer. Schmitz (1995) solved the Navier–Stokes equation by the implicit finite difference method. The result demonstrated that change in the Hagen–Poiseuille parabolic profile was proportional to the local transmembrane pressure (i.e., wall suction).

Kumar and Upadhay (2000) and Gupta et al. (2018) used a staggered grid approach using numerical backward Euler explicit integration to simulate ultrafiltration units with circular and rectangular cross-sections. It was observed that the staggered grid approach results are more accurate and required less computational time than other methods such as FDM, FVM, and FEM. Karode (2001) obtained pressure drop values in a channel with constant and varying permeation velocity at the membrane surface. It was shown that pressure drop is greater with the varying permeation velocity assumption. The pressure drop obtained from CFD predictions was verified with the analytical solutions for the constant permeation velocity case. Silva et al. (2011) observed the behavior of Newtonian and non-Newtonian fluid flow in the tubular membrane. A finite difference scheme for a staggered grid system with three convective terms was solved using a high order discretization to observe the cross-flow filtration process.

Extensive studies have illustrated the need for the readily available CFD tool to model and simulate the complete set of continuity and momentum equations of membrane filtration to predict the velocity, pressure, and concentration profile in membrane modules. Some commercial CFD software such as Ansys FLUENT™ (FVM based) and COMSOL™ (FEM based) have also been used for simulation to gain an insight phenomenal distribution of velocity, pressure, concentration, and wall shear stress on the membrane surface. However, the definition of the porous wall (domain boundary) in some software is misinterpreted/poorly defined by many. Some authors have made wrong assumptions, like membrane walls to be non-permeable.

Ghidossi et al. (2006) and Saber (2011) used the commercial Ansys fluent™ FVM package to determine the pressure drop and axial velocity profile along the hollow fiber membrane surface. Ahmad et al. (2005) studied concentration profiles, shear stress, and mass transfer coefficient in a membrane channel using the same commercial CFD package. Film theory, which is used as a boundary condition for a membrane wall, was integrated into the fluent with the help of a user-defined function (UDF). The boundary condition for a typical membrane wall obtained by the material balance of convective, diffusive, and permeate fluxes is given in Eqn. 3.19. The solution algorithm for fluent is shown in Figure 3.4

for $y = h$ and $0 < x < L$

$$D_{AB}\frac{\partial C_A}{\partial y} + J_v \times C_A = J_v \times C_{AP} \qquad (3.19)$$

Ahmad et al. (2005) integrated the membrane boundary condition in the CFD simulation code to predict the polarization profile in the membrane channel. They concluded that increasing shear stress, polarization profile decreases, and rise in TMP with decrease in Reynolds number would promote the formation of concentration polarisation. Ahmad and Lau (2007) further extended this work by including variable fluid properties and varying permeation velocities along the membrane channel. In wastewater treatment, membrane bioreactors (MBR) are becoming more efficient than conventional methods. Various researchers have investigated MBR reactors to

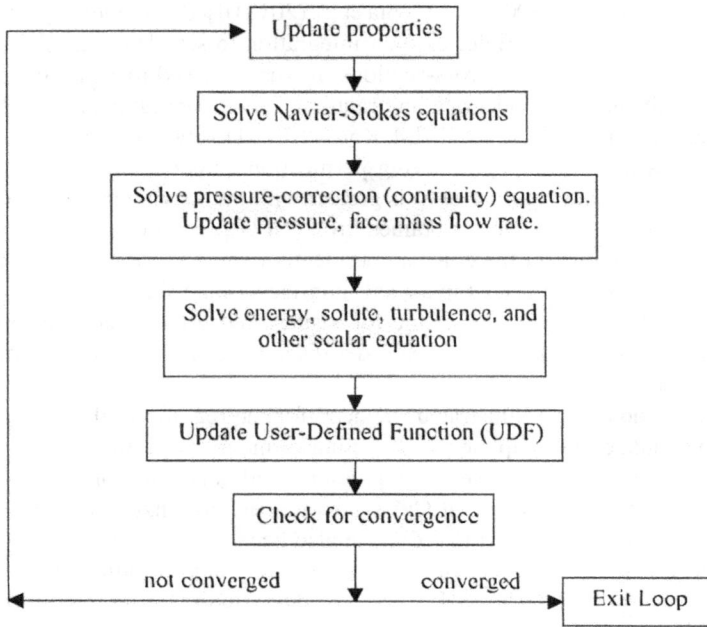

FIGURE 3.4   Simulation procedures using segregated solver in fluent (Ahmad et al., 2005).

study the effect of share stress, cross flow velocity, fluid flow patterns, and membrane fouling phenomena using the CFD tool. (Brannock et al., 2010; Ratkovich et al., 2011).

Membrane distillation (MD) is a growing membrane hybrid process technology and is currently being explored to be used for wastewater treatment. MD is found efficient in the treatment of water from the Arabian Gulf with a process achieving high flux and low energy demand (Alkhudhiri et al., 2013). The effect of different operating parameters on the MD has been investigated for its industrial applications. Shakaib et al. (2013) investigated the effect of spacer orientation, feed velocity, and temperature on MD modules. It was concluded that the advantage of the CFD method over conventional numerical techniques is that it can incorporate more factors such as local heat and mass transfer coefficients.

The hollow fiber module is extensively used in MD for wastewater treatment. Yang et al. (2012) studied the performance of hollow fiber modules on MD with the help of 3D-CFD simulation. They simulated straight fiber to wavy microstructure on hollow fiber surface to enhance the MD performance, which is only possible through visualization of the flow field and temperature profile obtained from CFD simulation. There is a high scope of MD and MBR in wastewater treatment with new strategies and designs for high flux and low polarization, and CFD tools seem to be an effective tool for this purpose.

3.9   CFD-BASED MEMBRANE PROCESSES IN WASTEWATER TREATMENT

The application of CFD in wastewater treatment with and without a membrane is widely acclaimed (Samstag et al., 2016). For wastewater treatment, membrane bioreactors (MBR) have been used for a long time (Naessens et al., 2012a; 2012b). With the help of CFD simulations, many advancements in the design of flow systems have been achieved. Amato and Wicks (2009) reported up-gradation and diagnostics of a dissolved air flotation (DAF) plant from 30 to 60 Ml/d using CFD analysis. The vector plots were used to visualize how changes made to the inclined baffle and tank depth can impact the subnatant water quality and turbidity. Saalbach and Hunze (2008) presented a CFD model to visualize flow-field in the hollow-fiber and flat sheet MBR. 3D-unsteady CFD simulation could successfully be applied to the nanofiber membrane contactor to estimate gas–liquid and gas–gas mass transfer coefficients in removing volatile organic compounds from wastewater (Dabiri, 2019). The steady-state simulation of vacuum-driven porous hollow-fiber membrane distillation process was made by Ghadiri et al. (2015). Removal of 1,1,1-trichloroethane (TCA) from an aqueous solution was simulated using the finite element numerical solution of the governing equations (Soltani et al., 2016). Barati et al. (2014) simulated toluene extraction from water using air as stripping gas in a porous hollow-fiber membrane contactor. The contactor was divided into three compartments: shell, porous membrane, and tube. Momentum and mass transfer equations were solved using the CFD tool in all zones of the contactor.

3.10   CONCLUSIONS

The use of membrane technology is vast in wastewater treatment. This chapter attempted to summarize the highlights of major membrane technology used for wastewater treatment. The membrane fouling and concertation polarization are the main limitations for every membrane process. There are many theoretical analyses based on mathematical models and empirical correlations, but none of them is completely satisfactory due to the limitations of the method used for solving the problem. This chapter discussed the importance of CFD, which is a combination of momentum balance and mass transfer equation for the analysis of permeate flux and fouling phenomena in the membrane module. The use of CFD for the modeling of the membrane allows the rigorous assessment of the fluid present at the adjacent layer of membrane surface by analyzing concentration polarization and fouling phenomena. The local information predicted by the CFD provides a better understanding of the process, making it easy to optimize the membrane module design for the required purpose.

Most of the reported simulation results of membrane-based systems are based on rigorous computer coding. With the advent of new commercially available CFD simulation tools (such as Ansys FLUENT™ and COMSOL™), researchers have started trying more complex problems such as the effect of concentration on the rheology of the permeate and retentate. However, some coding to create user-defined functions (UDF) is still needed to introduce such concentration-dependent property of the fluid (Ahmed et al., 2005; Awashthi and Kumar, 2017). On the other hand, since Navier–Stokes equation for simple channel geometry of membrane module are easier to solve, the same concentration-dependent rheology could be modeled with more excellent stability by numerical solution of the Navier–Stokes equation (Pak et al., 2008; Soltani et al., 2014; Gupta et al., 2017; 2018).

REFERENCES

Abdel-Fatah, MA. 2018. Nanofiltration systems and applications in wastewater treatment: Review article, Ain Shams Engineering Journal. 9: 3077–3092.

Ahmad, A.L. and Lau, K.K. 2007. Modeling, simulation, and experimental validation for aqueous solutions flowing in nanofiltration membrane channel. Ind. Eng. Chem. Res. 46: 1316–13.

Ahmad, A.L., Lau, K.K., Bakar, M.Z.A. and Shukor, S.R.A. 2005. Integrated CFD simulation of concentration polarization in narrow membrane channel. Comput. Chem. Eng. 29: 2087–2095.

Ahmed, F.N., 2013. Modified Spiegler-Kedem model to predict the rejection and flux of nanofiltration processes at high concentrations. Department of Civil Engineering, University of Ottawa, Canada.

Alkhudhiri, A., Darwish, N., Hilal, N. 2013. Produced water treatment: Application of air gap membrane distillation. Desalination, 309: 46–51.

Amato, T. and Wicks, J., 2009. The practical application of computational fluid dynamics to dissolved air flotation, water treatment plant operation, design and development. Journal of Water Supply: Research and Technology—AQUA, 58(1), pp.65–73.

Awasthi, KS, and Kumar, V. 2017. CFD Simulation of Membrane Separation Rheology, CHEMCON-2017, Haldia Institute of Technology, Paper ID MSO00479 (December 27–30, 2017).

Baker, R.W. 2002. Future directions of membrane gas separation technology, Ind. Eng. Chem. Res. 41: 1393–1411.

Baker, R.W. 2004. Membrane technology and applications, 2nd edition, John Wiley and Sons ltd, England.

Bakhshayeshi, M., Teella, A., Zhou, H., Olsen, C., Yuan, W. and Zydney, A.L. 2012. Development of an optimized dextran retention test for large pore size hollow fiber ultrafiltration membranes. J. Membrane Sci. 421–422: 32–38.

Barati, F., Ghadiri, M., Ghasemi, R. and Nobari, H.M., 2014. CFD simulation and modeling of membrane-assisted separation of organic compounds from wastewater. Chemical Engineering & Technology, 37(1), pp.81–86.

Belfort, G. 1988. Membrane modules: comparison of different configurations using fluid mechanics. J. Membrane Sci. 35: 245–270.

Berman, A.S., 1953. Laminar flow in channels with porous walls. J. Appl. Phys. 24: 1232–1235.

Bhattacharjee, C. and Bhattacharya, P.K. 1992. Prediction of limiting flux in ultrafiltration of kraft black liquor, J. Membrane Sci. 72: 137–147.

Bhattacharyya, D., Back, S.L., Kermode, R.I. and Roco, M.C. 1990. Prediction of concentration polarization and flux behavior in reverse osmosis by numerical analysis. J. Membrane Sci. 48: 231–262.

Brannock, M., Wang, Y., Leslie, G. 2010. Mixing characterization of full scale membrane bioreactors: CFD modeling with experimental validation. Water Res. 44: 3181–3191.

Cancilla, N., Gurreri, L., Marotta, G., Ciofalo, M., Cipollina, A., Tamburini, A. and Micale, G., 2022. A porous media CFD model for the simulation of hemodialysis in hollow fiber membrane modules. Journal of Membrane Science, p.120219.

Chang, In-S., Field, R. and Cui, Z. 2009. Limitations of resistance-in-series model for fouling analysis in membrane bioreactors: A cautionary note. Desalin. Water. Treat. 8:31–36.

Chatterjee, A., Ahluwalia, A., Senthilmurugan, S. and Gupta, S.K. 2004. Modeling of a radial flow hollow fiber module and estimation of model parameters using numerical techniques. J. Membrane Sci. 236: 1–16.

Cheng, T.-W., Yeh, H.O. M., and Wu, J.-G. 1998b. Permeate flux prediction by integral osmotic-pressure model for ultrafiltration of macromolecular solutions. J. Chin. Inst. Chem. Eng. 29: 193–199.

Childress, A.E. and Elimelech, M. 2000. Relating nanofiltration membrane performance to membrane charge (electrokinetic) characteristics. Environ. Sci. Technol. 34: 3710–3716.

Cohen, J.A. and Highsmith, S., 1997. An improved fit to website osmotic pressure data. Biophysical Journal, 73(3), p.1689.

Dabiri, E., Noori, M. and Zahmatkesh, S., 2019. Modeling and CFD simulation of volatile organic compounds removal from wastewater by membrane gas stripping using an electro-spun nanofiber membrane. Journal of Water Process Engineering, 30, p.100635.

Damak, K., Ayadi, A., Zeghmati, B. and Schmitz, P. 2004. A new Navier-Stokes and Darcy's law combined model for fluid flow in cross-flow filtration tubular membranes. Desalination. 161: 67–77.

de Pinho, M.N., Semião, V. and Geraldes, V. 2002. Integrated modeling of transport processes in fluid/nanofiltration membrane systems. J. Membrane Sci. 206: 189–200.

Ershad A., Mohammad R. M., Seyyed M.M. and Navid M. 2013. Experimental study and computational fluid dynamics simulation of a full-scale membrane bioreactor for municipal wastewater treatment application. Ind. Eng. Chem. Res. 52: 9930–9939.

Filipe, C.D.M. and Ghosh, R. 2005. Effects of protein-protein interaction in ultrafiltration based fractionation processes, Biotechnol. Bioeng. 91: 678–687.

Fimbres-Weihs, G.A. and Wiley, D.E., 2010. Review of 3D CFD modeling of flow and mass transfer in narrow spacer-filled channels in membrane modules. Chemical Engineering and Processing: Process Intensification, 49(7), pp.759–781.

Gartling, D.K., Hickox, C.E. and Givleer, R.C. 1997. Simulation of coupled viscous and porous flow problems. Int. J. Comput. Fluid Dyn. 7: 23–48.

Geraldes, V., Semião, V. and de Pinho, M.N. 2000. Numerical modelling of mass transfer in slits with semi-permeable membrane walls. Eng. Comput. 17:192–218.

Ghadiri, M., Asadollahzadeh, M. and Hemmati, A., 2015. CFD simulation for separation of ion from wastewater in a membrane contactor. Journal of Water Process Engineering, 6, pp.144–150.

Ghidossi, R., Daurelle, J.V., Veyret, D. and Moulin, P. 2006. Simplified CFD approach of a hollow fiber ultrafiltration system. Chem. Eng. J. 123: 117–125.

Ghosh, R. and Cui, Z.F. 2000a. Purification of lysozyme using ultrafiltration, Biotechnol. Bioeng. 68: 191–203.

Goksen, C. 2005. Development of a membrane based treatment scheme for water recovery from textile effluents. M.Sc. Thesis, Middle East Technical University.

Gupta R.R, Kumar, V. Chand, S. 2017. Flow behavior in weakly permeable microtube with varying viscosity near the wall. Polish Journal of Chemical Technology. Degruyter 19 (4):16—21.

Gupta, R.R., Kumar, V. and Chand, S., 2018. Effects of concentration dependent local resistances and viscosity on overall permeation characteristics of a hollow-fiber membrane: A simulation approach. Asia-Pacific Journal of Chemical Engineering, 13(5), p.e2236.

Hidane, T., Demura, M., Morisada, S., Ohto, K. and Kawakita, H., 2022. Mathematical analysis of cake layer formation in an ultrafiltration membrane of a phycobiliprotein-containing solution extracted from Nostoc commune. Biochemical Engineering Journal, p.108333.

Karode, S.K. 2001. A predictive model for ultrafiltration: combination of osmotic pressure model and irreversible thermodynamics. Separ. Sci. Technol. 36: 2659–2676.

Kavitha, E., Poonguzhali, E., Nanditha, D., Kapoor, A., Arthanareeswaran, G., and Prabhakar, S., 2022. Current status and future prospects of membrane separation processes for value recovery from wastewater. Chemosphere, 291, p.132690.

Keir, G., Jegatheesan, V. 2014. A review of computational fluid dynamics applications in pressure-driven membrane filtration. Rev Environ Sci. Biotechnol, 13: 183–201.

Kozinski, A.A., Schmidt, F.P. and Lightfoot, E.N., 1970. Velocity profiles in porous-walled ducts. Industrial & Engineering Chemistry Fundamentals, 9(3), pp.502–505.

Kumar, N.S.K., Yea, M.K. and Cheryan, M. 2004. Ultrafiltration of soy protein concentrate: Performance and modelling of spiral and tubular polymeric modules, J. Membrane Sci. 244: 235–242.

Kumar, V. and Upadhyay, S.N. 2000. Computer simulation of membrane processes: Ultrafiltration and dialysis units. Comput. Chem. Eng. 23: 1713–1724.

Kumar, V., Pandey, R.N. and Upadhyay, S.N. 1998. A closed form solution of convective mass transfer model for intracellular calcium response of endothelial cells, Math. Probl. Eng. 4: 437–459.

Kumar, V.D., Maity, D. and Purkait, M.K. 2012. Prediction of flux decline during membrane filtration of leather plant effluent, Int. J. Env. Waste Manage. 9: 123–140.

M. Shakaib, S.M.F. Hasani, M. Ehtesham-ul Haque, I. Ahmed, R.M. Yunus, 2013. A CFD study of heat transfer through spacer channels of membrane distillation modules. Desalin. Water Treat. 51: 3662–3674.

Marcos, B., Moresoli, C., Skorepova, J. and Vaughan, B. 2009. CFD modeling of a transient hollow fiber ultrafiltration system for protein concentration. J. Membrane Sci. 337: 136–144.

Mat, N.C. Lou, Y and Lipscomb, G.G. 2014. Hollow fiber membrane modules. Current Opinion in Chemical Engineering. 4:18–24.

Miroslav S., Bojan L. and Dejan R. 2003. Review of membrane contactors designs and applications of different modules in industry, ME Transition. 31: 91–98.

Mulder, M. 1991. Basic principles of membrane technology. Kluwer Academic. Dordrecht, Netherlands.

Naessens, W., Maere, T. and Nopens, I., 2012a. Critical review of membrane bioreactor models– Part 1: Biokinetic and filtration models. Bioresource Technology, 122, pp.95–106.

Naessens, W., Maere, T., Ratkovich, N., Vedantam, S. and Nopens, I., 2012b. Critical review of membrane bioreactor models–Part 2: Hydrodynamic and integrated models. Bioresource technology, 122, pp.107–118.

Nassehi, V. 1998. Modelling of combined Navier-Stokes and Darcy flows in cross flow membrane filtration. Chem. Eng. Sci. 53: 1253–1265.

Pak, A. Mohammadi, T. Hosseinalipour, S.M. and Allahdini, V. 2008. CFD modeling of porous membranes. Desalination. 222: 482–488.

Park, E. and Barnett, S.M., 2001. Oil/water separation using nanofiltration membrane technology. Separation Science and Technology, 36(7), pp.1527–1542.

Parsegian, V.A., Rand, R.P. and Rau, D.C., 1995. [3] Macromolecules and water: Probing with osmotic stress. In Methods in enzymology (Vol. 259, pp. 43–94). Academic Press.

Parvareh, A. Rahimi, M. Madaeni S.S. and Alsairafi, A.A. 2011. Experimental and CFD study on the role of fluid flow pattern on membrane permeate flux. Chin. J. Chem. Eng., 19: 18–25.

Purkait, M.K. Gupta, S. D. and De, S. 2004, Resistance in series model for micellar enhanced ultrafiltration of eosin dye, J. Colloid Interface Sci. 270: 496–506.

Ratkovich, N. Berube, P. R. Nopens, I. 2011. Assessment of mass transfer coefficients in coalescing slug flow in vertical pipes and applications to tubular airlift membrane bioreactors. Chem. Eng. Sci. 66: 1254–1268.

Saber, A. Seraji, M.T. Jahedi, J. and Hashib, M.A. 2011. CFD, Simulation of turbulence promoters in a tubular membrane channel. Desalination. 276: 191–198.

Samstag, R.W., Ducoste, J.J., Griborio, A., Nopens, I., Batstone, D.J., Wicks, J.D., Saunders, S., Wicklein, E.A., Kenny, G. and Laurent, J., 2016. CFD for wastewater treatment: An overview. Water Science and Technology, 74(3), pp.549–563.

Santoyo, A. B. Gomez-Carrasco, J.L. Gomez, G.E. Maximo Martin, M.F. and Hidalgo-Montesinos, A.M. 2004. Spiral-wound membrane reverse osmosis and the treatment of industrial effluents. Desalination. 160:151–158.

Sarkar, A. Moulik, S. Sarkar, D. Roy, A. and Bhattacharjee, C. 2012. Performance characterization and CFD analysis of a novel shear enhanced membrane module in ultrafiltration of bovine serum albumin (BSA). Desalination. 292: 53–63.

Sarkar, P. Datta, S. Bhattacharjee, C. Bhattacharya P.K. and Gupta, B.B. 2006. Performance study on ultrafiltration of Kraft Black Liquor and membrane characterization using Spiegler-Kedem model. Korean J. Chem. Eng. 23: 617–624.

Schausberger, P. Norazman, N. Li, H. Chen, V. and Friedl, A. 2009. Simulation of protein ultrafiltration using CFD: Comparison of concentration polarisation and fouling effects with filtration and protein adsorption experiments. J. Membrane Sci. 337: 1–8.

Schmitz, P. and Prat, M. 1995. 3-D laminar stationary flow over a porous surface with suction: Description at pore level. AIChE Journal. 41: 2212–2226.

Senthilmurugan, S. Ahluwalia, A. and Gupta, S.K. 2005. Modeling of a spiral-wound module and estimation of model parameters using numerical techniques. Desalination. 173: 269–286.

Shon, H. K., Phuntsho, S., Chaudhary, D.S., Vigneswaran, S., and Cho, J. 2013 Nanofiltration for water and wastewater treatment–a mini review. Drink. Water Eng. Sci., 6: 47–53.

Silva, J.M. Ferreira, V.G. and Fontes, S.R. 2011. An evaluation of three unwinding approximations for numerical modeling the flow in tubular membrane of Newtonian and non-Newtonian fluids. Appl. Math. Compu. 217: 7955–7965.

Singh, R. Nicholas, P. H. 2016. Introduction to Membrane Processes for Water Treatment, chapter 2, Emerging Membrane Technology for Sustainable Water Treatment.

Soltani, H., Pelalak, R., Heidari, Z., Ghadiri, M. and Shirazian, S., 2016. CFD simulation of transport phenomena in wastewater treatment via vacuum membrane distillation. Journal of Porous Media, 19(6).

Spiegler, K.S. and Kedem, O., 1966. Thermodynamics of hyperfiltration (reverse osmosis): criteria for efficient membranes. Desalination, 1(4), pp.311–326.

Sreenivas, K. Ragesh, P. Dasgupta, S. and De, S. 2002. Modeling of cross-flow osmotic pressure controlled membrane separation processes under turbulent flow. J. Membrane Sci., 201: 203–212.

Toh KY, Liang YY, Lau WJ, Fimbres Weihs GA. A review of CFD modelling and performance metrics for osmotic membrane processes. Membranes. 2020 Oct;10(10):285.

Yang, X. Yu, H. Wang, R. and Fane, A.G. 2012. Optimization of microstructured hollow fiber
    design for membrane distillation applications using CFD modeling. J. Membr. Sci. 421–
    422: 258–270.
Yuan, S. W. and Finkelstein, A. B. 1956. Laminar pipe flow with injection and suction through
    a porous wall. Trans, ASME J. Appl. Mech. 78: 719–724.
Zeman, L.J. and Zydney, A.L. 1996. Microfiltration and ultrafiltration: principles and
    applications. Marcel Dekker Inc., New York.

4 CFD or Modeling of Micromagnetofluidic/ Microfluidic Devices for Water Purification/ Water Treatment

Sudhanshu Kumar[1] and Krunal M. Gangawane[1,2]*
[1]Department of Chemical Engineering, National
Institute of Technology Rourkela, Rourkela, Odisha, India
[2]Department of Chemical Engineering, Indian
Institute of Technology Jodhpur, Jodhpur, Rajasthan, India
*Corresponding Author: Krunal M. Gangawane

CONTENTS

DOI: 10.1201/9781003325147-4

61

4.1 INTRODUCTION

Since a decade, substantial progress has been furnished in nanotechnology due to its pragmatic significance in numerous areas like sensors, microelectronics, medicine, wastewater treatment (WWT) (Luo et al. 2006; Salata 2004; Esakkimuthu, Sivakumar, and Akila 2014). Today CFD techniques are well pronounced in several textbooks (Anderson and Wendt 1995; Versteeg and Malalasekera 2007) and are usually applied in engineering science. CFD is one of the emerging fields in the WWT, by applications to unit processes. To identify and eliminate heavy metals from environment, reliable approaches are required. The primary innovation presented here is the integration of nanoparticle research with microfluidics and magnetohydrodynamics (MHD) for water treatment (Mitra and Chakraborty 2011). This chapter delivers an outline of CFD used to an extensive variety of processes in WWT, from hydraulic elements such as flow splitting to biological, chemical, and physical exercises such as anaerobic digestion, nutrient elimination, and suspended growth (Samstag et al. 2016). Microfluidics is the study of science and technology, which process (10^{-9} to 10^{-8} liters) quantities of fluids via channels with the dimension of ten to hundreds of micrometers (Whitesides 2006). Microfluidics is widely explored as an exclusive platform to produce nanoparticles with anticipated properties, i.e., morphology and size. CFD delivers an effective scheme to recognize the numerous effects on fluid flows without conducting complicated experiments. Benchmarking is an important segment in CFD and code validation based study (Horvat, Kljenak, and Marn 2001; Gupta and Kumar 2010). In contrast, the CFD approach emerges as a viable approach to overcome the constraints executed by experimental study (Soh, Yeoh, and Timchenko 2016). Another pragmatic significance of CFD is to analyze the droplet formation within microfluidic devices, which is important for droplet-based study. They assist as one of the valuable schemes for designing engineering apparatuses and systems in aerospace, turbo machinery, and several fields. Microfluidic technology has quickly evolved in the past decade because of its requirement, and the usage of this technique has augmented drastically in a few years. In this chapter, we emphasize the application of CFD as a potential tool for microfluidic devices and the specific application for water treatment and fluids' separation (Erickson 2005). It analyzes the features for monitoring of water, concentrating on several water resources (fresh, drinking, waste, and seawater) and their pragmatic application in commercial usage. Several organizations acknowledge water safety as the major challenge of the 21st century. The World Health Organization (WHO) lays the assessment of water quality as a worldwide importance, together with clean water for drinking, clean sanitization, and decent water supervision for human well-being. It covers a probable solution for the purification of water from heavy metals through magnetic based nanoparticles within microfluidic systems. Microfluidics evolved as a multidisciplinary arena, enticing numerous interests from commercial and scientific communities. The requirement for effective functioning control and hence more well-organized processes enhanced the advancement of microfluidic devices, which are broadly utilized in pragmatic applications involving micromixers, analytical separations, microreactors, bioassays, kinetic studies, and sample preparation steps. Microfluidics allows significant benefits to be found from microscale than macroscopic approaches. One can find these benefits with several instances in industrial processes. In this performance, the utmost significant component is the micromixer, while external magnetic fields achieve efficient mixing and particle driving. CFD are used to simulate water flow and nanoparticles. The two- and three-dimensional Navier–Stokes (NS) equations are solved for the flow field, while the magnetic nanoparticles' trajectories are simulated using a Lagrangian method (Mitra and Chakraborty 2011; Kefou et al. 2016). This technique is projected to flourish in chemical swiftness and improved water purification periods. The basic knowledge introduced in this chapter is of microfluidics for water purification. The multiphase flow encompassed of two or more immiscible fluids, which occurs in microfluidic devices, in which one phase wets the wall of the device and condenses the second one because of interfacial forces, taking place, in this way, discrete droplets. Additionally, the study of droplet-based microfluidics study earned enormous research attention from multi-disciplines because of its vantages and more controls. Abundant articles demonstrate its pragmatic applications in several fields (Song, Chen, and Ismagilov 2006; Günther and Jensen 2006; Marre and Jensen 2010; Teh et al. 2008). In the case of micro dimensions, surface forces govern over body forces necessitating peculiar consideration for a study concerning double-phase flows with free surfaces frequently determined by capillary forces. Distinctive flow conditions are capillary wicking or development of droplets characterized by small Weber (We) and Reynolds numbers (Re) (Glatzel et al. 2008). Another meticulousness in this study is the enhanced significance of diffusion for mixing phenomena. Meanwhile, in microfluidic based studies, Re are generally low, turbulence can be noticed, and mixing takes place by diffusion at small Peclet numbers (Pe) (Nguyen and Wu 2004). Monitoring of quality of drinking, fresh, waste, and sea water is of substantial significance to assure the security and well-being of human flora and fauna. Researchers are evolving vigorous water-monitoring microfluidic devices (Saez et al. 2021). Still, the conveyance of a costly, commercially accessible stage has not been attained yet, due to which it was focused with the help of CFD modeling for the treatment of wastewater. Traditional monitoring of water is primarily depended on laboratory apparatuses or sophisticated and costly hand-held investigations for on-site analysis, necessitating proficient peoples and time taking. Conventional microfluidics usages are air-filled and hydraulic regulators and pumps, which involve further external apparatus, diminishing the compactness by enhancing the value of the ultimate product.

4.2 PREVIOUS STUDIES

Numerous studies have been conducted to explore the detailed study of CFD modeling for microfluidics. Few studies deal with microfluidics for water purification/ water treatment (Table 4.1).

4.3 ABOUT MICROFLUIDICS

Microfluidics delivers a substantial opportunity to produce devices proficient in beating traditional chemical research and biomedical techniques. The work exercised at Stanford

TABLE 4.1
Previous Studies Associated with Microfluidics for Water Purification/Treatment

Sl No.	Source	Title	Remarks
1.	(Elshaw, Hassan, and Khan 2016)	CFD modeling and optimization of a wastewater treatment plant bioreactor-a case study.	The outcomes of the suspended solid concentration, while considering the flow as water, pointed out that the region is slightly above the suggested variety of 7 W/m³ and 12 W/m³. Through integrating into the CFD study oxygen tranport, nitrification, carbon oxidation, and denitrification, it will permit for a healthier apprehension of the impact of hydrodynamic execution and, so the ideal outline for submersible mixers.
2.	(Kefou et al. 2016)	Water purification in micromagnetofluidic devices: Mixing in MHD micromixers.	The optimum gradient magnetic field helped the navigation of molecules into a chosen trajectory far from the walls of the channel. Consequently, an improved and quicker mixing between molecules and fluid occurred. Since it was confident of its functioning, the model can be utilized to enhance the heavy metal detention technique further.
3.	(Santana et al. 2020)	A review based on the microfluidic device for water treatment processes	As established from the review, the escalation of liquid–liquid extraction benefits of small-size microfluidic devices.
4.	(Saez et al. 2021)	Microfluidics for water monitoring: A review	Augmentation in the number of journals each year and their quality shows the importance of microfluidic devices.
5.	(Qian and Lawal 2006)	Numerical study of liquid slugs and gas within a T-junction microchannel.	A T-junction-shaped empty microchannel of different cross-sectional breadth assisted as the model of microreactor, and a finite volume-based commercial CFD was chosen for the study.

University associated with microfluidics was first published in 1979. In the previous 35 years, it has progressed quickly from its grass roots in the microelectronics industries by the origin of miniaturized total analysis systems in various fields today. The amendment of present skills and the advancement of novel approaches has contributed to the formation of several devices, which permitted scientists to analyze the systems in an efficient, fast, and automated way in comparison to earlier (Convery and Gadegaard 2019; Nguyen, Wereley, and Shaegh 2019). Microfluidics is still an academic field rather than a useful commercial product. Microfluidics is frequently prefigured as a game changer in industry and life science-based research (Beebe, Mensing, and Walker 2002). Microfluidics scales between 100 nm and 100 μm; this revolution has seen an enormous interest in research nowadays, with several devices now proficient in beating their traditional ancestors besides, the advancement of novel devices has permitted for new workability and the examining of phenomena that were indescribable to macroscale devices (Qin et al. 1998). We have explained the physics behind the microscale to recognize the influences that govern the behavior of mixtures and liquids. These essences enlighten numerous of microfluidics benefits, like simple kinematics and faster reaction times. Microfluidics proposes a rising set of tools for deploying small amounts of fluids to control physical, biological, and chemical processes that are pertinent to sensing. The development of microfluidic devices, which are widely used in applications involving micromixers, sample preparation, microreactors, analytical separations, kinetic investigations, and bioassays, was sped up by the requirement for effective operating control and, consequently, more efficient processes. Although numerous developments in microfluidics took place by the 20th century, its origins began similarly to microelectronics. Due to its use in various disciplines, including chemistry, biology, and the physical sciences, microfluidics—the technique of fluid operation within channels has recently become a prominent new area of study. The progress of lab-on-a-chip (LOC) gadgets, anticipated to revolutionize chemistry and biology similarly as integrated circuits revolutionized computing proficiencies, is a major driving force in the field of microfluidic research. LOCs are microsystems proficient of incorporating whole chemical or biological laboratories within one chip, incorporation of microfluidic devices and active or passive mechanisms, like mixers, valves, filters, and several others. As a substitute, microfluidics has appeared as a potential tool that can substitute traditional analytical systems. The liquid–liquid extraction method via microfluidic is a well-explored technique.

Though, it has numerous advantages over limitations. It can be noted that in spite of the benefits in microfluidics, there are a few limitations, such as trouble in phase separation, limited approaches for study, and the partial flow range. Through the evolution of these novel approaches, the beginning of the century fetched an enormous rise in microfluidics, which contribute to the propagation of various microfluidic platforms with an extensive variety of functionalities. Plenty of these skills are defined briefly in this segment. Comprehensive reviews on the various arenas of microfluidics and their proficiencies can be observed elsewhere (droplet microfluidics (Shang, Cheng, and Zhao 2017), paper analytical devices (Yetisen, Akram, and Lowe 2013), open microfluidics (Kaigala, Lovchik, and Delamarche 2012), and organ-on-a-chip (Bhatia and Ingber 2014).

4.3.1 PHYSICS OF THE MICROFLUIDICS

To comprehend the assistance of miniaturised systems, it is significant to recognize the basics of flow on the scale and to know how it involves their behavior. Primarily, the relation of the inertial forces to viscous forces within the system is pronounced by Re, and it is presented in Eq. 4.1 (Reynolds 1883)

$$\mathrm{Re} = \frac{\rho v L}{\mu} \tag{4.1}$$

In the above equation, ρ denotes the density of the fluid, v is the characteristics velocity, and L is the characteristic length. Whereas μ is the dynamic viscosity. It is evident from the equation that as the characteristic length of the geometry or system is diminished, Re also gets reduced. As Re falls to < 2000, the system enters the laminar regime, conveys numerous variances over turbulent flow. Primarily, the laminar range is extremely foreseeable, meaning the mathematical modeling of systems is less rigorous.

Furthermore, the molecular passage in the laminar region varies from the turbulent as convective mixing is negligible, only diffusion that extends to highly predictable kinetics. In microfluidic, Re will be in the laminar range. Moreover, to Re, the Pe (Eq. 4.2) contributes a message on fluid species transport (Tabeling 2005).

$$\mathrm{Pe} = \frac{v L}{D} \tag{4.2}$$

where D symbolizes diffusion coefficient, Pe pronounces the relative importance of the advective to diffusive transfer ratio of molecules within the fluid. According to Eq. 4.2, dropping the size of a domain induces diminution in the Pe. This implies that the kinetics of the system is more predictable similar to the Re. Furthermore, the behavior of the surface of fluid varies from the macro to microscale. Surface tension designates a fluid's attraction to adapt its surface-to-air interface to decrease its free energy. Interfacial tension pronounces a similar phenomenon in binary immiscible fluids such as oil in water. In the area of droplet microfluidics, this phenomena has been very effectively used. These forces can be employed to transport fluids without the use of pumps since it dominates gravity on the microscale. Thirdly, as the characteristic dimensions are diminished, capillary forces overcome gravitational forces. The force that a fluid experiences when it passes over a porous substance or a narrow capillary is referred to as capillary forces. Finally, compared to conventional devices, microfluidic systems have substantially faster reaction times. This is caused by the systems' reduced dimensions, which result in shorter diffusion times for individual molecules. An estimation for diffusion time is revealed in Eq. 4.3

$$t = \frac{x^2}{2D} \tag{4.3}$$

where x pronounces the length travelled by single molecule of solute along one axis at time, t. From the aforementioned equation, it is evident that

$$t \propto x^2 \tag{4.4}$$

So, it is evident that as a system's characteristic magnitudes are reduced, less period is required for atoms to diffuse through the system, which results in quicker reaction times in microfluidic devices. Keeping in mind the above phenomena, researchers and scientists can regulate these effects in devices that can attain duties with great value to chemical and biological studies. Moreover, because of diminished size, systems devour fewer reagents compared to traditional fluidic platforms, inducing perfect devices when the price of chemicals is a matter. For instance, blood glucose meters require a few drops of blood, which render figures of blood glucose concentration within seconds, enabling the subject to display his status and follow their treatment programs from the comfort of their house.

4.3.2 WATER TREATMENT AND PURIFICATION

Microfluidic devices offer a technique to accomplish study in distant locations, empowering in situ investigation at the point of need. Microfluidic devices provide the advantages of reduced measurement durations, increased sensitivity, enhanced selectivity, and great reproducibility. Microfluidic platforms and devices have been delivered over the past several years for detecting contaminants in waste, sewage, fresh, drinking, and seawater. Devices for testing drinking purposes are typically made to look for bacteria and diseases. Devices for waste, fresh, sewage, and sea water, however, need to be able to be deployed for extended time periods in challenging environmental circumstances. Numerous reviews are accessible that contribute a broader outlook of microfluidic technology for ecological monitoring, but no one recapitulates the current technology for diverse water sources. Potable water contains very few pathogens, parasites, bacteria, and, thus vast volumes of water should be concentrated in order to meet the detection threshold for modern detection techniques. This is a restriction of the current technology, thus several techniques, such as force cytometry and spectroscopy (Sakamoto et al. 2007), laser scattering (Huang et al. 2018), electrophoresis, and electrochemical were acquired by microfluidic technology to dominate the very small number of pollutants or bacteria, which is to be examined in the model. An additional technique to ameliorate recognition is exploiting bacteria's intrinsic electrical properties by combining two electrical methods, dielectrophoresis (DEP) and impedance measurements. Incorporation with a microfluidic system can develop cost-effective, accurate, and simple portable platforms.

Alternative purification of water can be accomplished of sewage and wastewaters produced by numerous bases, where the concentration and kind of impurities observed in the existing high variability. Chemical pollutants and diseases that are harmful to the health of humans and ecosystems can have an impact on water quality. To identify and track the spread of the virus in a community, the levels of the SARS-COV-2 virus, for instance, are presently being measured in wastewater systems (Larsen and Wigginton 2020; Lancaster and Rhodes 2020). Because of this, society places high importance on routinely monitoring contaminants in wastewater systems using low-cost, straightforward methods. In this way, microfluidics has shown how to simultaneously measure and track a variety of contaminants in wastewater (Leelasattarathkul et al. 2007).

For example, copper is a common metal to find in the water. To assess Cu (II) in industrial effluent, a hybrid glass microfluidic system linked with an optical sensor was created. This invention appears to have a lot of potential for commercialization and application in systems for processing industrial effluent. It might serve as the foundation for creating an automated system for directly monitoring Cu (II) in sewage treatment facilities. Flowing streams, lakes, reservoirs, and groundwater are examples of freshwater bodies. For instance, water continues moving along the Earth's surfaces to form lakes or streams when rain occurs. It dissipates into the sky and absorbs into the aquifers that produce the public water supply. Sadly, public water sources can also be affected by chemicals or other contaminants found in the groundwater (Nguyen et al. 2019). It is crucial to monitor these waterways continuously to manage and safeguard water ecosystems. In this situation, sensors and devices based on microfluidics can monitor water quality and detect contaminants in fresh waters.

The balance between the quantity of substances like nitrate, ammonium, nitrite, and phosphate, which are referred to as nutrients, and salinity determines the quality of seawater to a large extent. Nutrient concentrations that are too high could have negative effects on marine flora and fauna. In addition, it is crucial to monitor the water quality and stop any decline that could be detrimental to marine ecology. These factors include temperature, pH, and conductivity. Microfluidic devices could be used in commercial operations like desalination facilities. In this regard, integrated microfluidic sensors have been created that can quickly monitor the conductivity, temperature, and salinity of the water extracted from a pilot-scale desalination plant (Kim et al. 2013).

4.4 MACROSCOPIC GOVERNING EQUATION

The equations introduced in this study are pronounced below. For the incompressible flow is conveyed as (Bruus 2014; Sattari, Hanafizadeh, and Hoorfar 2020):

$$\nabla.\vec{V} = 0 \qquad (4.5)$$

$$\rho_f\left(\frac{\partial \vec{V}}{\partial t} + \left(\vec{V}.\nabla\right)\vec{V}\right) = -\nabla P + \eta\nabla^2\vec{V} + \vec{F}_d \qquad (4.6)$$

In which $\vec{F}_d$ is the drag force exercised by the magnetic particle, $\vec{V}$ is the velocity vector of fluids, P is the pressure, and η represents the dynamic viscosity of the fluid.

The simplified species transfer equation is pronounced as

$$\frac{\partial C}{\partial t} + \vec{V}.\nabla C = D\nabla^2 C \qquad (4.7)$$

where C is the concentration, D is the constant diffusion coefficient.

Newton's second law governs the motion of magnetic particles:

$$m_p \frac{\partial \vec{v}}{\partial t} = \vec{F}_d + m_p \vec{g} + \vec{F}_m \qquad (4.8)$$

where $m_p \vec{g}$ is denoted as the gravity force of every magnetic particle, which is equivalent to 8.1×10^{-3} pN, and that can be ignored in assessment to the large magnetic actuation force, the drag force ($\vec{F}_d$), which is estimated as,

$$\vec{F}_d = C_D \rho_f \left(\vec{V} - \vec{v} \right) \left| \vec{V} - \vec{v} \right| \frac{A_p}{2} \qquad (4.9)$$

where, the drag coefficient (C_D) banks on the Rayleigh number (Ra), $\vec{V}$ and $\vec{v}$ represents the velocity vector of the flow and the magnetic particles, correspondingly. A_p designates the cross-sectional area of the particles. $\vec{F}_m$ is the magnetic force exercised on the magnetic particles and can be evaluated with the help of the following equation:

$$\vec{F}_m = V_p \left(\overline{M} \cdot \nabla \right) \vec{B} \qquad (4.10)$$

In the above equation $\vec{F}_m$ represents magnetic force, V_p denotes the volume of the particles, $\overline{M}$ designates magnetization of the particle, and $\vec{B}$ defines flux density produced from the external electromagnets. Because of the presence of an electromagnet, the magnetic effect in the y-direction is more robust than in the x and z-directions. So, the component B_x and B_z can be untreated, and the consistent magnetic actuation force F_m is substantial in the y-direction only.

4.5   BOUNDARY CONDITIONS

Boundary conditions (BCs) in hydrodynamics are the constraints to boundary value study in CFD. These BCs consist of the inlet, outlet, axisymmetric, symmetric, wall, constant pressure, and periodic or cyclic BCs. Transient-based studies involve initial conditions where initial values of flow variables are quantified at nodes within the domain (Ashgriz and Mostaghimi 2002). BCs are set as the flow arrives at the microchannel at an inlet with constant velocity $V_0=1$ mm/s. No-slip BCs are implemented on the walls of the microchannel. At the exit, constant pressure is functional. The flow field and concentration are set as the initial conditions for the study. CFD technique arises as a potential numerical tool to beat the limitations enforced by experiments (Table 4.2).

TABLE 4.2
Numerical Methods and Boundary Conditions Used in the CFD Simulations (Chaves et al. 2020)

Discretization Method	The Finite Volume Method
Pressure-velocity coupling algorithm:	Pressure-Implicit with Splitting of Operators (PISO)
Pressure interpolation method (IM):	PRESTO
Spatial IM:	Second order upwind scheme
Temporal IM:	Euler method
Wall boundary condition (BC):	No-slip BC
Inlets BC:	Velocity inlet BC
Outlet BC:	Outflow BC
Interface Interpolation:	Geometric reconstruction method
Body force (BF) formulation:	Implicit BF

4.6 CASE STUDY VALIDATIONS

Optimum CFD study and proposal of a microfluidic device that evades prolonged trial and error microfabrication save considerable technical and financial resources. Hence, numerical studies to visualize particle movement have been endeavored, leading to an ideal design. Fluent version 6.1.22 from Fluent Inc. was utilized for this case study. Fluent's software packages include the Fluent solver for geometry creation and meshing and T Grid for generating volume meshes from existing boundary meshes. Gambit offers a standard set of CAD operations as well as integrated scripting capabilities for quick geometry development. Gambit offers structured, unstructured, hybrid, non-conformal, and structured meshes, including quadrilateral in 2D and hexahedron, prism in 3D, and triangle in 2D and tetrahedron. The Cooper scheme or manually creating structured meshes both enable quick meshing with high-quality grids.

After providing suitable mesh size, Fluent is utilized for the study setup, the initialization, and the post-processing of the outcomes. All GUI actions are programmed by the software during simulation setup, but it is important to be aware that not all options are available through the GUI. There are several specific functions that can only be used by entering them at the command line level, which can make it challenging to identify the right parameters. Most of the problems involving structured and unstructured meshes that are mentioned in this work can be addressed by Fluent. Fluent offers four distinct VOF formulations for the VOF model: the geo-reconstruct method, the explicit Euler formulation, the donor-acceptor method, and the implicit formulation.

4.6.1 RESULTS OF THE SIMULATION FOR THE SINGLE-OUTLET MICROFLUIDIC DEVICE

Using the solver FLUENT 6.3.26, the flow and concentration profile of the particles in the microchannel were analyzed. To investigate the flow pattern of the dispersed particles in the presence of a non-uniform electric field (EF), the simulation of the microchannel with dual inlets and single outlet was performed. The inlet velocities at inlet 1 and inlet 2 are 10^{-5} m/s and 5×10^{-5} m/s, respectively. Three pairs of electrodes were placed in the wall at a length to length ratio of 1:10 each. A distance of 1000 m was maintained between the adjacent large electrodes. The microchannel measured 149 m and 0.73 cm in width and length, respectively. Figure 4.1(a) demonstrates the potential distribution (PD) inside the microchannel after 5 V and 200 V of applied voltage are delivered to the smaller and larger electrodes, respectively. This information was acquired numerically by figuring out the Laplace equation. Figure 4.1(b) depicts the distribution profile of the electric field, with the small electrodes exhibiting the greatest field strength. The arrows show the orientation of the EF lines as they transition from a higher to a lower potential.

Similarly, the distribution of $\nabla E2$ is illustrated for a functional PD of 195 V among the opposite electrodes.

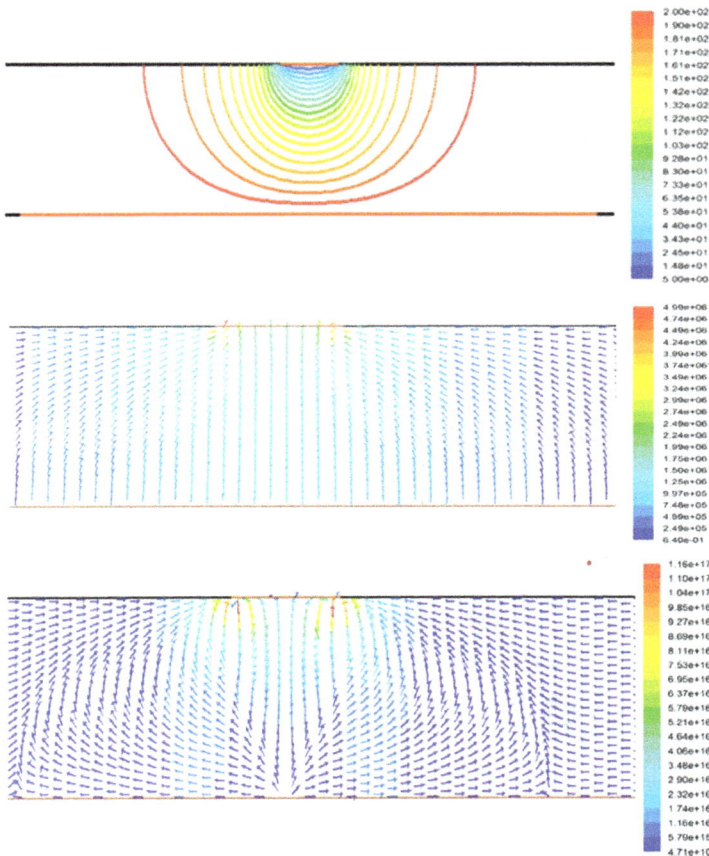

FIGURE 4.1 Contours of (I) voltage distribution, (II) electric field distribution, (III) ∇E^2 for a potential difference of 195 V functional to the electrode 1:10 (Dash et al. 2016).

4.6.2 DESIGN AND PROCESS PARAMETER OPTIMIZATION

Optimizing several parameters is necessary when designing a microfluidic device for effective separation via DEP. To obtain a strong DEP force, for instance, a low applied voltage to the electrodes combined with a high electric field gradient is desirable. Particles of various sizes must exit through different outlets; hence the several electrodes and the layout of the outlets must be optimized. To enhance separation efficacy with the previously optimised parameters, the water's velocity must also be optimized.

4.6.3 PROBLEM DESCRIPTION

The studies are carried out in a rotating reference frame with stationary channel walls, i.e., a reference frame that rotates at the disk's speed. All simulation tools must correctly translate the boundary conditions into the rotating frame for this analysis. This three-dimensional study was conducted for a channel of 10 mm in length, directing in the radial direction at a distance of 20 mm from the midpoint of rotation, as revealed in Figure 4.2. The channel's symmetry plane is in the middle of the cross-section, which has the dimensions $a = b = 200$ m.

To reduce the computational cost, the numerical study was accomplished on the structured grid as displayed in Figures 4.2(III) and (IV), which feats the planar symmetry of the study. The complete channel contains 1010 evenly spaced nodes throughout its length, 16 and 30 nodes along its height, and 391,608 cells as a result. The stationary numerical study was executed at non-slip BC on the walls and constant pressure of 0 Pa at the entry respectively exit.

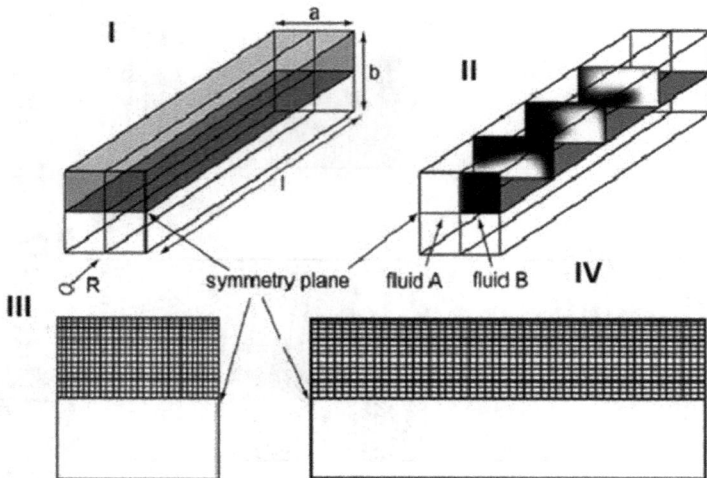

FIGURE 4.2 (I) Square-shaped channel with radial edges. (II) Patterns of the fluid A and B through the channel. (III) The cross-section mesh of the channel (IV) Mesh in the direction of the channel (Glatzel et al. 2008).

4.6.4 MEASUREMENT OF FLOW RATES AND PRESSURE DROP THROUGHOUT THE CHANNEL

The variation of the pressure in the channel's center as a function of radial position is an effective quantitative metric to evaluate the simulation results for this issue. The pressure drop along the spinning channel is not linear, in contrast to pressure-driven flow. Further missing from the center of rotation, the pressure rises as a result of the increasing centrifugal force. This results in the renowned convex pressure profile, as displayed in Figure 4.3.

Although in this instance neither experimental data nor analytical solutions are available for validation, it appears that CFX produces the most logical outcome. According to the fundamental physical interpretation of this model, the pressure curve begins at $p = 0$ Pa at the entry, drops in a convex manner, and ends at $p = 0$ Pa at the exit. Remarkably, Fluent fails to preserve $P = 0$ Pa at the entry; moreover, to this offset, Fluent offers a solution with a surprising overshoot inside the initial 200 μm length, which cannot be enlightened even qualitatively by the fundamental physics. Such an overshoot also exists in the consequences developed by CFDACE+; still, it is significantly slighter, and the BC's solution is at minimum steady at the entry and exit.

All of results are quite constant within 20 Pa, with the exception of the variations at the entrance and outflow. Even at the far end of the channel, consistency increases, indicating that the variances among the codes may decrease as one moves far from the center of rotation. This has to be verified by a more in-depth investigation, though. The leaving global flow rate was ascertained to be around 16 μl s^{-1} taking to $Re = 90$ via codes apart from for Flow-3D, wherever the flow frequency of 3 μl^{-1} was estimated.

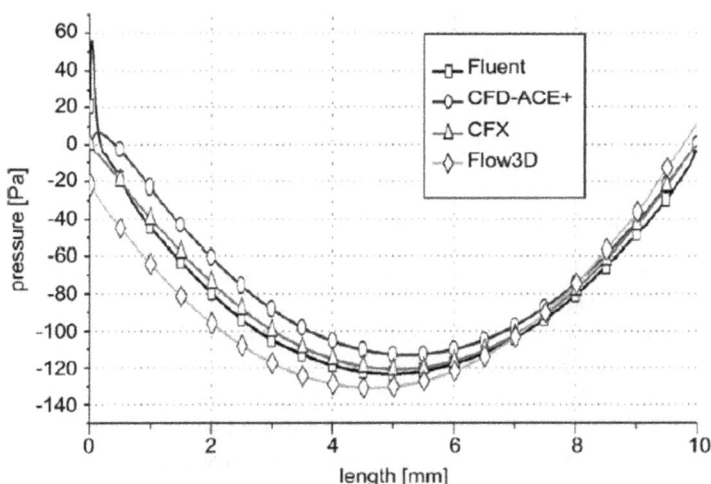

FIGURE 4.3 Pressure distribution toward the mid of the channel from entry to exit (Glatzel et al. 2008).

4.6.5 COMPUTATIONAL TIME

The calculation of computational time furnished in the subsequent subdivision only reproduces the execution of the earlier considered software tools pertaining to the previously studied benchmark problems that are obtainable in this study. It should be noted that either program could recreate any issues more quickly. For most software tools, it is not optimal to build up the given issues with a structured grid and predetermined times step to get maximum computing performance. The technique of tracking the streamline using a scalar value was selected for mixing study and numerical diffusion analysis. Usually, a steady-state solution is needed for such issues. Since Flow-3D had a transient solver, tracking the scalar value over the entire geometry takes a lengthy time to simulate. A similar circumstance occurred in the Fluent simulations conducted for the rotating channel. As opposed to a typical stationary solution, a manual iterative solution was required in this case, which significantly lengthens the overall computing time. As a result, these outcomes cannot be directly compared to those of the other techniques. The calculation time is quite important for free surface studies simulated using the VOF approach. The VOF algorithm, in particular the surface reconstruction, significantly increases the computing load to start. Second, a very narrow time step (on the order of $10E^{-6}$ s to $10E^{-5}$ s) is necessary to resolve the capillary oscillation of the free surface in order to reach a convergent solution, which is frequently far too small for the typically essential study time frames. As a result, the simulation times take much longer than they would for stationary study, making the efficiency of computational speed very interesting. A strong physical comprehension of the problem at hand is necessary to minimize the model size in order to maximize computational speed. The only ways to speed up the simulation process inside the solver setup are to employ automated, non-iterative time steps or parallel processing if that option is accessible. Unstructured grids sometimes speed up processing by reducing the number of computational cells needed to solve a given problem. This can be very helpful for complex geometries, but when combined with VOF, it becomes more challenging because surface reconstruction methods on unstructured grids are more expensive and complex than on structured grids. In order to maintain the results as comparable as feasible, all of these possibilities for increasing computational speed were not purposely used in the prior study.

4.6.6 COMPUTATION TIME: T-SHAPED CHANNEL

The volume of fluid (VOF) approach is expensive in the form of computing time. The T-shape channel research, however, was estimated very quickly using 3D. This is due to the fact that Flow-3D is not constrained to continuous time stepping, whereas the other programs were. The left programs took five to ten times more computing time, whereas Fluent is roughly three times slower than Flow-3D. CPU running times for the incredibly difficult Top Spot challenge ranged from 463 h to 634 h. The difference in simulation times is roughly 1.5, employing CFX constituted the quickest scheme.

Given that there is no geometric surface reconstruction, this is not surprising. Fluent is comparable in speed to CFD-ACE+ but a little slower. A non-iterative or adaptive time-stepping approach could reduce the computation time for this problem.

It is stated that Fluent's latest version (6.2.16), in particular, the time-stepping technique, can speed up simulations by five times.

4.6.7 COMPUTATION TIME: ROTATING CHANNEL, RECOMBINING, AND SPLIT STRUCTURE

There are no specific recommendations to speed up the simulation process for divide and recombine structures because the tools address such situations by sensible simulation periods. Fluent is one of the quickest tools, clocking in at 1.25 hours, followed by CFD-ACE+ at 3 hours. While Flow-3D requires roughly 56 times more factor than Fluent's fastest tool, CFX and Flow-3D were the moderate approaches for computing the recombine and split structure. This is due to Flow-3D's inability to offer the ability to run a steady-state simulation. Since the simulation tools had to alter the BC and forces within the rotating system, the rotating channel-based challenge was more difficult. It was easy to set up CFD-ACE+; however, one must exercise caution when applying boundary conditions, as inlets are appropriately placed within the system. By the help of CFD-ACE+ settings, which is faster than another technique, is the tool that might be used in this case study.

Fluent requires more time to simulate the rotating channel since, as previously mentioned, an iterative procedure rather than a single solver run was required to arrive at a valid solution. Flow-3D experiences long simulation periods due to the transient simulation requirement and the spinning channel's high aspect ratio.

REFERENCES

Anderson, John David, and John Wendt. 1995. *Computational fluid dynamics*. Vol. 206: Springer.

Ashgriz, Nasser, and Javad Mostaghimi. 2002. An introduction to computational fluid dynamics. *Fluid Flow Handbook* 1:1–49.

Beebe, David J, Glennys A Mensing, and Glenn M Walker. 2002. Physics and applications of microfluidics in biology. *Annual Review of Biomedical Engineering* 4 (1):261–286.

Bhatia, Sangeeta N, and Donald E Ingber. 2014. Microfluidic organs-on-chips. *Nature Biotechnology* 32 (8):760–772.

Bruus, Henrik. 2014. Governing equations in microfluidics.

Chaves, IL, LC Duarte, WKT Coltro, and DA Santos. 2020. Droplet length and generation rate investigation inside microfluidic devices by means of CFD simulations and experiments. *Chemical Engineering Research and Design* 161:260–270.

Convery, Neil, and Nikolaj Gadegaard. 2019. 30 years of microfluidics. *Micro and Nano Engineering* 2:76–91.

Elshaw, Andrew, NMS Hassan, and MMK Khan. 2016. CFD modelling and optimisation of a Waste Water Treatment Plant Bioreactor-a case study. Paper read at 2016 3rd Asia-Pacific World Congress on Computer Science and Engineering (APWC on CSE).

Erickson, David. 2005. Towards numerical prototyping of labs-on-chip: modeling for integrated microfluidic devices. *Microfluidics and Nanofluidics* 1 (4):301–318.

Esakkimuthu, T, D Sivakumar, and S Akila. 2014. Application of nanoparticles in wastewater treatment. *Pollut. Res* 33 (03):567–571.

Glatzel, Thomas, Christian Litterst, Claudio Cupelli, et al. 2008. Computational fluid dynamics (CFD) software tools for microfluidic applications–A case study. *Computers & Fluids* 37 (3):218–235.

Günther, Axel, and Klavs F Jensen. 2006. Multiphase microfluidics: from flow characteristics to chemical and materials synthesis. *Lab on a Chip* 6 (12):1487–1503.

Gupta, Amit, and Ranganathan Kumar. 2010. Flow regime transition at high capillary numbers in a microfluidic T-junction: Viscosity contrast and geometry effect. *Physics of Fluids* 22 (12):122001.

Horvat, Andrej, Ivo Kljenak, and Jure Marn. 2001. On incompressible buoyancy flow benchmarking. *Numerical Heat Transfer: Part B: Fundamentals* 39 (1):61–78.

Huang, Wei, Limei Yang, Gang Yang, and Feng Li. 2018. Microfluidic multi-angle laser scattering system for rapid and label-free detection of waterborne parasites. *Biomedical optics express* 9 (4):1520–1530.

Kaigala, Govind V, Robert D Lovchik, and Emmanuel Delamarche. 2012. Microfluidics in the "open space" for performing localized chemistry on biological interfaces. *Angewandte Chemie International Edition* 51 (45):11224–11240.

Kefou, Nikolitsa, Evangelos Karvelas, Konstantinos Karamanos, Theodoros Karakasidis, and Ioannis E Sarris. 2016. Water purification in micromagnetofluidic devices: mixing in MHD micromixers. *Procedia Engineering* 162:593–600.

Kim, Myounggon, Wooyeol Choi, Hyuk Lim, and Sung Yang. 2013. Integrated microfluidic-based sensor module for real-time measurement of temperature, conductivity, and salinity to monitor reverse osmosis. *Desalination* 317:166–174.

Lancaster, Kari, and Tim Rhodes. 2020. Wastewater monitoring of SARS-CoV-2: lessons from illicit drug policy. *The Lancet Gastroenterology & Hepatology* 5 (7):641–642.

Larsen, David A, and Krista R Wigginton. 2020. Tracking COVID-19 with wastewater. *Nature Biotechnology* 38 (10):1151–1153.

Leelasattarathkul, Tapparath, Saisunee Liawruangrath, Mongkon Rayanakorn, Boonsom Liawruangrath, Winai Oungpipat, and Napaporn Youngvises. 2007. Greener analytical method for the determination of copper (II) in wastewater by micro flow system with optical sensor. *Talanta* 72 (1):126–131.

Luo, Xiliang, Aoife Morrin, Anthony J Killard, and Malcolm R Smyth. 2006. Application of nanoparticles in electrochemical sensors and biosensors. *Electroanalysis: An International Journal Devoted to Fundamental and Practical Aspects of Electroanalysis* 18 (4):319–326.

Marre, Samuel, and Klavs F Jensen. 2010. Synthesis of micro and nanostructures in microfluidic systems. *Chemical Society Reviews* 39 (3):1183–1202.

Mitra, Sushanta K, and Suman Chakraborty. 2011. *Microfluidics and Nanofluidics Handbook: Chemistry, physics, and life science principles*: CRC Press.

Nguyen, Nam-Trung, Steven T Wereley, and Seyed Ali Mousavi Shaegh. 2019. *Fundamentals and applications of microfluidics*: Artech house.

Nguyen, Nam-Trung, and Zhigang Wu. 2004. Micromixers—a review. *Journal of Micromechanics and Microengineering* 15 (2):R1.

Nguyen, Tien Thanh, Manh Truong Dang, Anh Vu Luong, Alan Wee-Chung Liew, Tiancai Liang, and John McCall. 2019. Multi-label classification via incremental clustering on an evolving data stream. *Pattern Recognition* 95:96–113.

Qian, Dongying, and Adeniyi Lawal. 2006. Numerical study on gas and liquid slugs for Taylor flow in a T-junction microchannel. *Chemical Engineering Science* 61 (23):7609–7625.

Qin, Dong, Younan Xia, John A Rogers, Rebecca J Jackman, Xiao-Mei Zhao, and George M Whitesides. 1998. Microfabrication, microstructures and microsystems. In *Microsystem technology in chemistry and life science*: Springer.

Reynolds, Osborne. 1883. XXIX. An experimental investigation of the circumstances which determine whether the motion of water shall be direct or sinuous, and of the law of resistance in parallel channels. *Philosophical Transactions of the Royal society of London* (174):935–982.

Saez, Janire, Raquel Catalan-Carrio, Róisín M Owens, Lourdes Basabe-Desmonts, and Fernando Benito-Lopez. 2021. Microfluidics and materials for smart water monitoring: A review. *Analytica Chimica Acta* 1186:338392.

Sakamoto, Chieko, Nobuyasu Yamaguchi, Masumi Yamada, Hiroyasu Nagase, Minoru Seki, and Masao Nasu. 2007. Rapid quantification of bacterial cells in potable water using a simplified microfluidic device. *Journal of Microbiological Methods* 68 (3):643–647.

Salata, Oleg V. 2004. Applications of nanoparticles in biology and medicine. *Journal of Nanobiotechnology* 2 (1):1–6.

Samstag, R W[iiii], J J[iiii] Ducoste, A[iiii] Griborio, et al. 2016. CFD for wastewater treatment: an overview. *Water Science and Technology* 74 (3):549–563.

Santana, HS, JL Silva, B Aghel, and J Ortega-Casanova. 2020. Review on microfluidic device applications for fluids separation and water treatment processes. *SN Applied Sciences* 2 (3):1–19.

Sattari, Amirmohammad, Pedram Hanafizadeh, and Mina Hoorfar. 2020. Multiphase flow in microfluidics: From droplets and bubbles to the encapsulated structures. *Advances in Colloid and Interface Science* 282:102208.

Shang, Luoran, Yao Cheng, and Yuanjin Zhao. 2017. Emerging droplet microfluidics. *Chemical Reviews* 117 (12):7964–8040.

Soh, Gim Yau, Guan Heng Yeoh, and Victoria Timchenko. 2016. Improved volume-of-fluid (VOF) model for predictions of velocity fields and droplet lengths in microchannels. *Flow Measurement and Instrumentation* 51:105–115.

Song, H, DL Chen, and RF Ismagilov. 2006. Angew,"Reactions in droplets in microflulidic channels" Chemie, int. *Ed* 45:7336–7356.

Tabeling, Patrick. 2005. *Introduction to microfluidics*: Oxford University Press on Demand.

Teh, Shia-Yen, Robert Lin, Lung-Hsin Hung, and Abraham P Lee. 2008. Droplet microfluidics. *Lab on a Chip* 8 (2):198–220.

Versteeg, Henk Kaarle, and Weeratunge Malalasekera. 2007. *An introduction to computational fluid dynamics: the finite volume method*: Pearson Education.

Whitesides, George M. 2006. The origins and the future of microfluidics. *Nature* 442 (7101):368–373.

Yetisen, Ali Kemal, Muhammad Safwan Akram, and Christopher R Lowe. 2013. based microfluidic point-of-care diagnostic devices. *Lab on a Chip* 13 (12):2210–2251.

5 Coagulant Treatment

G.S.N.V.K.S.N. Swamy Undi[1] and
Chandra Shekar Bestha[2]
[1]Air Ok Technologies Private Limited, Research Park,
Indian Institute of Technology, Chennai, Tamil Nadu, India
[2]Department of Chemical Engineering, National Insititute of
Technology Calicut, Kozhikode, Kerala, India

CONTENTS

5.1　INTRODUCTION

Water treatment is becoming increasingly significant. The coagulation process is the most important step in the water treatment process because it makes it easier to remove colloids and tiny particles by feeding coagulants. Coagulation is a procedure that is often carried out by introducing outside substances to the water. However, flocculation is a physical process without a charge-neutralizing component, whereas coagulation is a chemical process that incorporates charge neutralization. Four different mechanisms—ionic layer compression, adsorption, charge neutralization, trapping in a flocculent mass, and adsorption and interparticle bridging—generally work together to cause chemical coagulation. Chemicals known as coagulant agents promote the coagulation of liquids into solids. When certain chemicals are introduced to raw water that contains slowly settling or non-settleable particles, the result is known as coagulation. Adding substances that encourage the clumping of fines into bigger flocs so they may be more readily removed from the water is known as coagulation-flocculation. The smaller particles eventually coalesce into bigger, heavier flocs that are collected by sedimentation and filtering. Flash mixing is the process of combining the coagulant chemical with untreated raw water. The main goal of the flash-mix method is to quickly mix and distribute the coagulant chemical uniformly throughout the water column. The production of extremely small particles is one of the first outcomes of the process, which is completed in a relatively little period. Coagulation-flocculation is a pre- or postprocessing technique that may be applied to water and wastewater treatment procedures, including filtration and sedimentation. Iron and aluminium-based coagulants are the most used, although titanium and zirconium have also shown promise as coagulants.

The effectiveness of the coagulation-flocculation process depends on a number of factors, some of which include the following:

- Type of coagulant used

Metallic solids release hydrogen ions and neutralizes alkalinity. The reduction of alkalinity causes a drop in pH value and can affect the coagulation.

- Coagulant dosage

The dosage of the coagulant depends on the turbidity of raw water, low turbidity and high dosages prompt sweep coagulation. The predominant mechanism is high turbidity and low dosages charge neutralization and adsorption.

- Final pH

Coagulation is most effective between pH levels 5 to 7.5. Coagulants like ferric chlorides, ferric sulfates, alum, and synthetic polymers.

- Coagulant feed concentration
- Type and dosage of chemical additives other than primary coagulant (e.g., polymers)
- Sequence of chemical addition and time lag between dosing points
- Intensity and duration of mixing at rapid mix stage
- Type of rapid mix device
- Velocity gradients were applied during the flocculation stage.
- Flocculator retention time
- Type of stirring device used
- Flocculator geometry.

The typical flow sheet of a water treatment plant (WTP) includes sequential stages (Figure 5.1)

Coagulants for water treatment fall into two main categories: organic and inorganic.

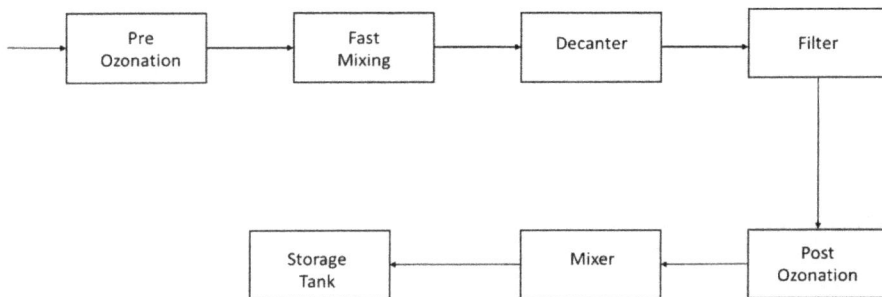

FIGURE 5.1 Flowsheet of water treatment plant.

5.1.1 ORGANIC COAGULATION

They are typically employed to produce sludge and to separate solids from liquids. There are two different kinds of organic compounds used in water treatment; the first is polyamines, which are the most popular organic coagulants. Simply neutralizing the charges of the particles to allow them to join is the most efficient way to clean wastewater and high-turbidity raw water. Second, tannins and melamine formaldehyde, which are utilized to coagulate colloidal particulates in water, are organic wastewater compounds. Due to its excellent ability to absorb organic pollutants like oil and grease, it is particularly useful in treating toxic sludge.

5.1.2 INORGANIC COAGULATION

Although inorganic coagulants may be used in a wide range of water treatment processes and are frequently more cost-effective than their organic equivalents, they are acidic. When organic coagulants are ineffective, inorganic coagulants can be employed to treat low turbidity water since they are particularly successful at doing so. The majority of inorganic coagulants are made of aluminium or iron. While alternative options for wastewater treatment include aluminium chloride, poly aluminium chloride, aluminium chlorohydrate, ferric and ferrous sulphate, and ferric chloride, aluminium sulphate is the chemical that is most commonly employed on a global scale.

Water undergoes a number of reactions when metallic salts are added to it, such as ferric or aluminium sulphate. To produce a precipitate, there must be enough chemicals added to the water to surpass the metal hydroxide's maximum solubility (floc). The floc will afterward adsorb on turbidity (water particles).

Polyelectrolytes are monomeric compounds with ionizable groups that are often used to treat water. Anionic polyelectrolytes are polymers with negatively charged particles, whereas cationic polyelectrolytes are polymers with positively charged groups on the monomeric units. Nonionic polymers are those that do not include ionizable groups.

5.1.3   How Does It Work?

The coagulation hypothesis is intricate. Insoluble flocs are created during coagulation, a physical and chemical interaction between the alkalinity of the water and the coagulant injected into it. After adding a chemical to the water to be treated, the hydrolysis species that prevail depends on the water's pH. Lower pH levels often favor positively charged species because they will interact with negatively charged colloids and particles to produce insoluble flocs, which filter out pollutants from the water.

Between pH 5 and pH 7, on average, is the ideal pH range for coagulation. Because coagulants typically react with the alkalinity in water, pH levels must be kept constant. The water's alkalinity serves as a buffer, reducing pH shifts and assisting in the chemical precipitation process. Unless it is very low, alkalinity in source water is often not a problem. Alkalinity can be raised by introducing lime or soda ash.

Positively charged molecules make up coagulants, which neutralize the electrical charge of particles and weaken the magnetic forces that hold colloids apart. Colloid particle precipitates are created when inorganic coagulation is introduced to wastewater and the cationic metal ion is neutralized. These particles trap pollutants and purify the water. This is the "sweep-floc" mechanism, which refers to the total volume of sludge that needs to be treated and removed. Similar steps are followed in the organic chemical coagulation process, however, an amine group linked to the coagulant molecule most frequently provides the positive charge rather than metal.

5.1.4   Methods of Mixing

Several methods can be used to mix the chemicals with the water to be treated:

- Hydraulic mixing using flow energy in the system
- Mechanical mixing
- Diffusers and grid systems
- Pumped blenders

In systems where the water is moving at a fast enough rate to create turbulence in the water, hydraulic mixing using baffles or throttling valves works effectively. Water flow turbulence combines chemicals with water.

Paddles, turbines, and propellers are frequently employed for coagulation. Although a mechanical mixer is adaptable and dependable, it uses a lot of power to combine the coagulant and water.

Pumped blenders can also be used in coagulation facilities for mixing. Through a diffuser in a pipe, the coagulant is immediately introduced to the water being treated. It permits the coagulant to spread out quickly and does not cause a major head loss. Compared to mechanical mixers of equivalent rating, electrical mixers use a great deal less energy.

The rapid mixing unit design parameters include mixing time t, and velocity gradient G.

$$G = (\frac{p}{V\mu})^{\frac{1}{2}}$$

(5.1)

where
G = velocity gradient, s^{-1}
P = power input, W
V = volume of mixing height basin, m^3
μ = viscosity, N.s/ m^2

Coagulation basins, which are rectangular tanks with mixing equipment, are the perfect mixing apparatus for chemical coagulants. If the flow velocity is high enough to cause turbulence, mixing may also take place in the influent channel or a pipeline leading to the flocculation basin. The design of the flash-mixing mechanism is also heavily influenced by the geometry of the basin.

5.2 COAGULANT DOSAGE TECHNIQUES

5.2.1 Tests to Determine Coagulation Dose

1. Jar Test
2. Microscale Dewatering Tests
3. Streaming Current Detector

1. Jar Test
Jar testing simulates a full-scale water treatment process and gives system operators a good sense of how a treatment chemical will behave and work with a specific kind of raw water. Jar testing allows system operators to choose which treatment chemical will perform best with the raw water in their system since it simulates full-scale operation. Jar tests make it possible to choose and dose chemical coagulants appropriately in order to remove suspended matter and contaminants from wastewater treatment facilities. Chemical coagulants are chosen and dosed based on the outcomes of the laboratory jar test, which simulates full-scale operation in water treatment facilities. The most popular coagulants are lime (calcium hydroxide), Alu (aluminium sulphate), and iron salts (ferric or ferrous).

The sample is swirled continually to allow for the observation of floc creation, development, and settlement, much as would happen in a real water treatment facility. The operator then conducts a series of experiments to evaluate the outcomes of various dosages of flocculation agents at various pH ranges in order to establish the ideal floc size.

Analytical procedures steps:
Take a few tall, 105-mm-diameter glass beakers (jars) each holding 1000 ml.
Several beakers (jars) are filled to the same level with each water sample, which is typically raw water. 600 cc of wastewater samples plus coagulant are added to each beaker of the water sample to treat it with a varied dosage of the chemical.
Jar testing often involves many beakers being tested simultaneously, with the findings from the first test directing the selection of parameter levels in the subsequent tests. Placing the paddles in the center of the sample and swirling it vigorously

for 120 seconds at a speed of 120 rpm. 30 rpm for up to 25 minutes of moderate-speed flocculation.

The results of the jar test can be evaluated based on different criteria:

After 5 minutes of sedimentation, the initial examination of the data. Utilize a turbidimeter to measure the supernatant's residual turbidity following a specified sedimentation period. Measuring the electro-kinetic potential of suspended particles on a sample that was obtained just after chemicals had been added and mixed. By using standardized membrane filters under pressure, clarified water's filterability is assessed. The degree of filter blockage brought on by lingering, unsettled suspended debris is correlated with the reduction in water flow.

Seasonally (temperature), monthly, weekly, daily, or anytime a chemical is changed, new pumps, quick mix motors, floc motors, or chemical feeders are added, and jar testing should be performed. Jar testing is not needed to be done on a regular basis, but the more it is done, the more efficiently the plant will run. The secret to running the plant more effectively is optimization.

Limitations of jar test
A jar test's applicability is constrained by a number of drawbacks. Significant amounts of water/wastewater samples and experimental time are needed to evaluate potential coagulants or flocculants, which limits the number of replicated experiments that may be carried out. Additionally, the outcomes of jar test experiments analyses are sometimes just semi-quantitative. In addition, there are many other chemical coagulants and flocculants that are regarded more of an "art" than a science.

2. Microscale Dewatering Tests

LaRue's invention, the microscale flocculation test, shrinks the scale of traditional jar tests to that of a multi-well standard microplate, resulting in advantages from the smaller sample volume and increased parallelization. This technique is also amenable to quantitative dewatering metrics, such as capillary suction time.

The streaming current detector is a computerized technique for calculating coagulant dosage (SCD). When the charges are balanced, the SCD monitors the surface charge of the particles and indicates zero streaming currents (cationic coagulants neutralize anionic colloids). The ideal coagulant dosage is at this value (0).

3. Streaming Current Detector

The streaming current detector is a computerized technique for calculating coagulant dosage (SCD). When the charges are balanced, the SCD monitors the surface charge of the particles and indicates zero streaming currents (cationic coagulants neutralize anionic colloids). The ideal dosage of the coagulant is at this value (0).

5.3 MODELS OF COAGULANT DOSAGE

Three categories of models can be made. The first kind relies only on raw water quality to forecast coagulant dose. Most models are trained using historical raw water quality data in order to forecast coagulant dose. The kinetics of the coagulation process is not covered by this kind of model. The process and inverse process of coagulation are models that are addressed and created to predict treated water quality from raw water and coagulant dosage (Maier, 2004; Specht, 1991; Zounemat-Kermani & Teshnehlab, 2008).

The process of model development algorithms is applied to create the model given (Figure 5.2).

The value of model predictions involves comparing many dynamic models to static models for determining the dose of coagulants. Analyze every model, taking into account the seasons, using at least a one-year data set. Put together a regression model that makes use of three sensors and accurately predicts the doses. To enhance the accuracy of estimating the coagulant dose in water treatment facilities, a number of strategies, including outlier removal and down-sampling of data, were explored (WTP). Streaming current detectors (SCD) are the primary means of automated coagulant control because they measure the charge on tiny, suspended particles in a liquid (Sibiya, 2014). Sensors are included because they are used to monitor processes. To determine the ideal coagulant dose, regression models, multilayer perceptron (MLP), and neural network models were applied. Temperature, pH, turbidity, color, ultraviolet absorbance at a wavelength of 254 [nm] (UVA-254), the concentration of dissolved organic carbon (DOC), and alkalinity were the inputs utilized to create the raw water quality parameters for the MLP model. The MLP model performs best in terms of treated water turbidity and coagulant dose. The output of the conventional MLP with h neurons and one hidden layer is provided by

$$y(k) = z(k)^T w_O + b_O$$

$$\text{where } z(k) = g(u(k)^T W_I + b_I)$$

here $y(k)$ = is the model output,

$z(k)$ = is the hidden layer output vector,

$u(k)$ = is the kth sample of input vector,

g = is tangent sigmoid,

W_I = is the matrix of weights connecting the D inputs to h hidden layer nodes,

and b_I = is the vector of hidden layer node biases.

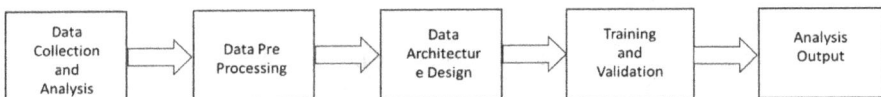

FIGURE 5.2 Model development process.

The output weights that connect the hidden neurons to the output neuron and the output bias are represented by w_O and b_O, respectively.

Using sensor data of raw water parameters, adaptive neuro-fuzzy inference system (ANFIS) models were assessed to forecast the coagulant dose. With the input water characteristics of turbidity, conductivity, temperature, dissolved oxygen (DO), UVA-254, and pH obtained from process sensors, an ANFIS model was trained to anticipate the coagulant dose. Regression methods offer the benefit of requiring fewer input variables to execute dose in a WTP with fewer sensors. Having fewer input variables lowers the cost of implementation. To ascertain the optimum regression model to anticipate the best trade-off between complexity and prediction outcomes. Modeling includes operational data, which includes details on the parameters values from both raw and coagulated water as well as the coagulant dose values (established by jar tests). The models were trained using a few years' worth of data, and the next few years' worth of forecasts, encompassing the wet and dry seasons, were used as a test set.

Elman recurrent network (ERN), distributed time-lagged feed-forward network (DTLFN), focused time-lagged feed-forward network (FTLFN), and nonlinear autoregressive with exogenous inputs are examples of dynamic neural network (DNN) models (NARX). The FTLFN model belongs to a broad category of dynamic networks known as focused networks, where the dynamics only manifest at the input layer. For a FTLFN model to be fully specified, a number of important decisions must be made (Gomes et al., 2015). The algorithm used to train the model parameters W_I, b_I, w_O, and b_O is among them. Others include the maximum time lag p_x, the number of hidden neurons h, the activation function of hidden layer nodes g, and the number of hidden neurons. The size of $u(k)$ should be carefully chosen because it grows proportionately to the greatest latency. The model input's dimensionality is equal to

$$D = N p_x + 1$$

N = number of input variables

Large D values can increase the training time and even lead to poor generalization in the prediction results.

The FTLNF model is frequently used in a variety of different fields, including stock market forecasting and the prediction of total suspended solids (TSS) in water treatment facilities.

The DTLFN model was used for time series prediction and traffic forecasting. This model contains inputs uj and one hidden layer (k), The DTLFN model's output is determined by the input vector of $u(k) = [\varphi\,(x_1(k), p_x) \dots \varphi(x_N(k)\,p_x)]^T$. where each $uj(k)$ equals the result of an application of a finite impulse response (FIR) filter to an input variable (xj) (k).

where each $uj(k)$ represents the result of an application of a finite impulse response (FIR) to an input variable (xj) (k). The result of applying a FIR filter on the input variable xj (k), corresponds to a weighted sum of actual and past values of this variable

$$\varphi\big(x_1(k), px\big) = \sum_{i=0}^{px} c_{ij} x_j (k-j)$$

CIJ is the coefficient of filter FIR of each input variable.

A group of context units in the ERN model are in charge of remembering the previous states of hidden units. The ERN controls the network by combining the input at the moment and the hidden units' prior states. The ERN model's ability to produce time-varying patterns is its most significant benefit. It is used in a variety of situations, such as transmission-line failure prediction. The hidden layer outputs, z, and $x(k)$, together with a collection of their previous values, make up the input $u(k)$ of the ERN model (k). The number of hidden neurons, or h, and the number of previous values of the input variables, or p_x, are the model parameters for the ERN model, as the number of past values of the hidden layer output p_z.

The nonlinear autoregressive with exogenous input (NARX) model is a common method for identifying linear black-box models. Recurrent neural network connections from the output layers are fed back to the input layer in the NARX model. The NARX model calls for the definition of the number of hidden neurons h, the number of previous input values p_x, the number of past output values p_y, and the activation function of hidden layer nodes g. The p_x and p_y parameters influence the NARX model's input dimensionality D, which is $D = N p_x + p_y + 1$. The input variables for dose prediction are the pH in raw and coagulated water as well as the turbidity in the coagulated water. These DNN models have the same one hidden layer and h neurons as the conventional multilayer perceptron model (MLP) (Table 5.1).

5.3.1 Evaluation Criteria of Model Performance

The model with the lowest prediction errors is the best. Numerous statistical indicators that may be classified using absolute and relative approaches can be used to measure prediction error. Mean absolute error (MAE) allows users to compare models directly on an absolute scale, while mean absolute percentage error (MAPE), which evaluates error proportionally, makes it simple for users to calculate the percentage difference between the observed and projected values for comparison (Gagnon

TABLE 5.1
Inputs and Output of Models

| Model | Raw Water Quality | | | | Settled Water | |
	Turbidity	pH	Temperature	Conductivity	Turbidity	Coagulant Dosage
ANFISA	I	I	I	I		O
MLP	I	I	I	I	O	I
DNN	I	I	I	I	I	O
UNIT	NTU		°C	µs/cm	NTU	ppm
I, input, O, output						

et al., 1997). While coefficients of determination ($R2$) show how many variations in projected values may be explained by the model, coefficients of correlation (R) are used to gauge how closely the anticipated outcomes and real values are related. The normalized root means square error (RMSE) is a useful metric for assessing the estimating abilities of various models.

Since it only reveals the degree of linear relationships between variables, the correlation coefficient is often not a good indicator for assessing model effectiveness when used alone. In other words, the R-value might be large if the observed and modeled values diverge much. In order to compare the best-fit model with each method's top performance on the test data set, external verification indices were used. The criterion involved comparing the regression line between the actual observed and anticipated values with the ideal fit line, which has a slope of one and an intercept of zero. If the model performs well, the absolute values are smaller than 0.1. When the discrepancies between the anticipated and observed values are minimal, the two values are comparable.

5.4   CASE STUDIES

5.4.1   BAN SONG WATER TREATMENT PLANT

The Ban Song WTP in Changwon City, South Korea, provided the information used in this study. The Korea Water Resources Corporation oversees the WTP (Kwater). The Nakdong River serves as the source of raw water for the WTP, which has a daily capacity of 120,000 gallons of water purification. Coagulation, flocculation, sedimentation, filtering, and chlorination injection are common processes. Additionally, they began pre-ozonation operations in May 2014. Since 2014, the coagulant used by Ban Song WTP is polyaluminum hydrogen chloride silicate (PAHCS) (Park & Choi, 2018).

Online analyzers for pH, turbidity, temperature, and conductivity can track water quality in real time. These aspects of the water quality become crucial operating variables for the coagulation process. The relationship between pH and floc features and coagulant solubility. Turbidity in low TOC water regulates the formation of excellent floc by destabilizing suspended colloids. The temperature has an impact on the effectiveness of floc formation rate and primary particle removal. Conductivity, which is a property of ionized substances in water, has an impact on dissolved solids. Each process' residence duration varies depending on the flow rate parameters.

Most of the time, operators use a reference table and data on water quality to establish the target value of the coagulant dose. Hourly data from the database server for the whole year of 2014 were pulled for this study. As a result, 8760 records in total are obtained. For WTP modeling, data from a whole cycle should be prepared.

5.4.1.1   Variation of Water Quality and Coagulant Dosage

Because of seasonal variations in the Nakdong River, Ban Song WTP's raw and settled water quality changed dramatically in 2014. Ninety percent of raw water's turbidity levels are below 19.77 NTU. But during the rainy season, from August to

September, it rises to about 440 NTU with very steep slopes (Kim & Parnichkun, 2017). Throughout the whole season, the temperature varies between 2.5 °C to 30 °C, with some months seeing lows of sub 5 °C. The pH ranges from 6.76 to 8.85; during the winter's dry season and the brief summer, it reaches over pH of 8.0. The average conductivity is 299, and the range is 110 to 540. The conductivity curve demonstrates that it significantly dropped during the wet season. The high variance of settled water turbidity is caused by the significant variation of raw water turbidity. Its surplus fluctuates above 1 NTU during the summer.

5.4.2 BEARSPAW WATER TREATMENT PLANT (WTP)

In order to forecast the coagulant dose in this article, annual water quality records from Alberta Environment for the Bearspaw Water Treatment Plant (WTP) in Alberta (Canada) were utilized. Support vector machine (SVM) and *K*-nearest neighbors (KNN) were studied (Orhan et al., 2011). Daily checks were made of the pH, temperature, and turbidity of the source waters. Color, conductivity, total dissolved solids, alkalinity, and other characteristics of source water quality are also measured on a regular basis. Additionally, the daily coagulant dose was noted. A WTP's operators choose the dose of the coagulant (alum or polyaluminum chloride, or PAC) depending on the variations in the source water's temperature, pH, and turbidity. For the Bearspaw WTP, water quality data were gathered between 2008 and 2010 and used in modeling. 966 group records in all were gathered.

5.4.2.1 Data Neutralization and Training

The normalized approach and the random sample methodology are used for the selection of training data and testing data in order to increase the dependability of the SVM regression model. To prevent solution divergence due to the orders of magnitude discrepancies between the parameters, the given data set is first normalized to the range [0, 1] (Zhang et al., 2013).

Seventy-five percent of the samples are utilized for model training, while the remaining 25 percent are used for model testing. All data records are split into two groups. Which data record was utilized for training and testing was determined using the random sample approach. The coagulant dose is then predicted using the input data from testing data and the trained SVM regression model. The performance of the SVM regression model is then assessed by contrasting the projected results with the actual dose measurements.

5.4.3 THE CASE STUDY OF THE CEARA WATER TREATMENT PLANT, BRAZIL

The approach is used to choose the optimal input variables, define the parameters for the DNN models, and determine the architecture for each DNN model. By reducing the mean square error (MSE) of all network output samples, the DNN model was trained to determine the values for the network parameters. Utilizing the Levenberg–Marquardt method, the models were trained. Any neural network model must have a certain amount of hidden neurons, or *h*, in order to be applied properly. In general, improved performance during the training phase is encouraged by a bigger number of neurons in the hidden layer. A neural network model cannot accurately describe the system being modeled if it is trained with a too-low number of hidden neurons, a condition known as under-fitting. The number of neurons that accurately reflects the system and function adequately during training must be chosen.

5.4.3.1 Variable Selection and Criteria

Each DNN model's parameters, referred to here as must be chosen. A vast input space where the values are connected might lengthen the training process and potentially provide predictions with poor generalization. The performance criteria for the evaluation of a DNN model trained with the parameter will be the trade-off between complexity and accuracy when using a subset of input variables. A nonlinear autoregressive with exogenous input (NARX) model with three input variables—the pH of raw and coagulated water and the turbidity of the coagulated water—contains the optimum answer.

5.5   RESULTS AND DISCUSSION

5.5.1   Coagulant Dosage Prediction Model using ANNs and ANFIS

The MLP, GRNN, and ANFIS methods are used in this section to determine the coagulant dosage in real time based on the fluctuating raw water quality. A method that is frequently used in model development is cross-validation. It may choose when to terminate training and gives comparative data on how well various models generalize data. The entire set of data is divided into training, validation, and test subsets (Zhang & Stanley, 1997, 1999; Zhu et al., 2015).

Analyzing correlations between raw water quality inputs and PAHCS dosage output is necessary to ascertain the impact of inputs on the model for forecasting PAHCS dosage. A statistical measure known as Pearson's correlation coefficient is used to examine the link between input parameters and PAHCS dose (R). Turbidity and PAHCS do have the highest correlation coefficient when compared to the other inputs, making them the most pertinent to the result. In terms of absolute value, the coefficients between pH, temperature, conductivity, and PAHCS dose range from 0.135 to 0.247. Additionally, there is a correlation between the input parameters as well. Eight alternative input combinations, always including turbidity, are chosen for the development of the MLP, GRNN, and ANFIS models based on the findings of the Pearson analysis. In order to identify the optimal model that produces the test data set's greatest correlation coefficient and lowest RMSE, each MLP model was evaluated using a range of hidden layer neuron numbers from the smallest to the largest. For two layers of hidden and output, respectively, two tangent sigmoid and linear functions are used. The LM algorithm trains the MLP models. To avoid overfitting, the cross-validation criteria is used as a termination condition. By adjusting the spread parameter's range, the spread parameter constant for GRNN is examined to find its optimal value. The spread setting that produces the greatest performance on the test data set with the lowest RMSE and highest correlation coefficient is chosen. Subtractive clustering is used in the creation of ANFIS models. The type of the membership functions is chosen to be Gaussian. By using a first-order polynomial and the Sugeno technique, fuzzy inference systems are created.

The outcome shows that the MLP and ANFIS models satisfy every evaluation requirement, indicating that the prediction performances of the two models are very accurate and trustworthy. The GRNN model nevertheless satisfies the other two prerequisites even when and requirements are not met. The most accurate prediction is made by the ANFIS model, which has a 10% lower relative peak error than the MLP model. In comparison to the value of the GRNN model, the measured value and the predicted values of MLP and ANFIS are more comparable to the measured PAHCS dosage value, and the prediction value of ANFIS also more closely tracks the observed value. The raw water models of MLP, ANFIS, and GRNN are assessed using the MAPE, RMSE, R, and $R2$ criteria in order to examine each best-fit model's resistance to turbidity fluctuation. The MLP model makes the most accurate predictions throughout the whole turbidity range, whereas the GRNN model performs the worst overall (Heddam et al., 2012). However, it is clear that each model fits the data well, in particular, turbidity zones. ANFIS is more superior than GRNN and MLP for turbidity zones with raw water turbidity ≤ 10 NTU. Both the MLP and ANFIS models have promising prediction accuracy. MLP, however, outperforms ANFIS in terms of accuracy in high turbidity zones. This supports research findings showing that neuro-fuzzy systems are effective at processing vast amounts of noisy data (Zounemat-Kermani & Teshnehlab, 2008). Performance enhancement may be attained by merging each best-fit model in a certain turbidity zone, depending on the efficacy of each approach.

5.5.2 BEARSPAW WTP SVM MODEL WITH KERNEL FUNCTIONS

The training and testing results from the application of the SVR models with three-order polynomial and radial basis functions for the Bearspaw WTP were chosen at random. When used with RBF kernel functions and third-order polynomials, the SVR exhibit good results. For the SVR model, sigmoid and linear functions were also tried. While testing the SVR model with sigmoid kernel function has a significantly lower coefficient of determination, it has a larger mean error and coefficient of determination. As a result, the RBF is chosen as the kernel function for the medium water treatment plant's training and testing.

Performances of the KNN method and the SVR model with RBF kernel function are compared. The same samples used in the SVR model and the prediction results produced using the KNN. KNN's performance for the WTPS is comparable to that of the SVR model in terms of results. The KNN method is more appropriate for online prediction since it can be implemented quickly and does not require a model training procedure. The operator of a WTP must quickly determine the coagulant dosage depending on changes in source water quality. Finding comparable cases in the database and then assigning the average to the query point is the best approach to accomplish this. The source water quality might vary significantly depending on certain circumstances, such as runoff and snowmelt.

The support vector machine regress models and the KNN method perform well for large and medium-sized WTPs, as demonstrated by the coagulant dose in big, medium, and small-sized WTPs. For small-sized WTPs, however, the KNN method outperforms the outcomes from SVR models. The quality of the data records affects how well the SVR and KNN models work. Only in a case-by-case scenario is it simple to declare which machine learning approach is superior to others. The ultimate choice may be made by contrasting various learning techniques like support vector machine and KNN.

5.5.3 THE EXPERIMENTAL RESULTS FOR THE PAC AND AS DOSAGES PREDICTION AT CEARA

The PAC and AS doses were predicted using two different methods. The pH and turbidity in the raw and coagulated water were the variables utilized for modeling the PAC and AS doses. A number of DNN models have been assessed using various combinations of the input variables. Based on the Cp statistics, the ideal design and the appropriate input variables for each model were chosen. Only three sensors—the pH of the raw and coagulated water as well as the turbidity of the coagulated water—are required to determine the PAC dose. Only the pH of the input water is shared by all DNN models; the other variables are distinct for each model.

5.6 CONCLUSIONS

A decision support tool to calculate the dose of coagulant using full-scale data was built using the MLP, GRNN, ANFIS with subtractive clustering, and K-means clustering approaches. To determine the ideal design, several input combinations and a wide variety of algorithmic parameters were simulated. On the models, the cross-validation approach was used. In-depth model analyzes were conducted using reliable statistical metrics. The outcomes showed that all of these models were highly accurate at predicting PAHCS dose. The modeling factors that worked best were temperature and turbidity. When additional input variables were added, the models' accuracy increased. All four of the water quality measures displayed complicated nonlinear features during the coagulation process. Future settling water quality as determined by raw water quality and coagulant dose was not predicted by the NARX-ANFIS model with time series feedback. The NARX-ANFIS, which uses current data on the quality of settling water and raw water turbidity to anticipate future turbidity, may be another excellent option for real-time coagulant dose management.

The quality of the data records affects how well the SVR and KNN models work. Only in a case-by-case scenario is it simple to declare which machine learning approach is more superior than others. The ultimate choice may be made by contrasting various learning techniques like support vector machine and KNN.

The doses of PAC and AS in a genuine WTP may be accurately predicted by a regression model using a limited number of input variables. For the PAC and

AS dosage predictions were attained using the NARX model. The pH in raw and coagulated water, as well as the turbidity in the coagulated water, are three sensors that are required for forecasts. With excellent coefficients of determination, the NARX model has a strong capacity to forecast PAC values accurately and AS dose levels in real time.

REFERENCES

Gagnon, C., Grandjean, B. P. A., & Thibault, J. (1997). Modelling of coagulant dosage in a water treatment plant. *Artificial Intelligence in Engineering*, *11*(4), 401–404. https://doi.org/10.1016/S0954-1810(97)00010-1

Gomes, L. S., Souza, F. A. A., Pontes, R. S. T., Fernandes Neto, T. R., & Araújo, R. A. M. (2015). Coagulant dosage determination in a water treatment plant using dynamic neural network models. *International Journal of Computational Intelligence and Applications*, *14*(03), 1550013. https://doi.org/10.1142/S1469026815500133

Heddam, S., Bermad, A., & Dechemi, N. (2012). ANFIS-based modelling for coagulant dosage in drinking water treatment plant: a case study. *Environmental Monitoring and Assessment*, *184*(4), 1953–1971. https://doi.org/10.1007/s10661-011-2091-x

Kim, C. M., & Parnichkun, M. (2017). MLP, ANFIS, and GRNN based real-time coagulant dosage determination and accuracy comparison using full-scale data of a water treatment plant. *Journal of Water Supply: Research and Technology–Aqua*, *66*(1), 49–61. https://doi.org/10.2166/aqua.2016.022

Maier, H. (2004). Use of artificial neural networks for predicting optimal alum doses and treated water quality parameters. *Environmental Modelling & Software*, *19*(5), 485–494. https://doi.org/10.1016/S1364-8152(03)00163-4

Orhan, U., Hekim, M., & Ozer, M. (2011). EEG signals classification using the K-means clustering and a multilayer perceptron neural network model. *Expert Systems with Applications*, *38*(10), 13475–13481. https://doi.org/10.1016/j.eswa.2011.04.149

Park, E., & Choi, H. (2018). The case study on wireless lan design technique for Bansong purification plant using network integrated management system and security switch. *Journal of the Korean Society of Water and Wastewater*, *32*(4), 309–315. https://doi.org/10.11001/jksww.2018.32.4.309

Sibiya, S. M. (2014). Evaluation of the Streaming Current Detector (SCD) for Coagulation Control. *Procedia Engineering*, *70*, 1211–1220. https://doi.org/10.1016/j.proeng.2014.02.134

Specht, D. F. (1991). A general regression neural network. *IEEE Transactions on Neural Networks*, *2*(6), 568–576. https://doi.org/10.1109/72.97934

Zhang, K., Achari, G., Li, H., Zargar, A., & Sadiq, R. (2013). Machine learning approaches to predict coagulant dosage in water treatment plants. *International Journal of System Assurance Engineering and Management*, *4*(2), 205–214. https://doi.org/10.1007/s13198-013-0166-5

Zhang, Q., & Stanley, S. J. (1997). Forecasting raw-water quality parameters for the North Saskatchewan River by neural network modeling. *Water Research*, *31*(9), 2340–2350. https://doi.org/10.1016/S0043-1354(97)00072-9

Zhang, Q., & Stanley, S. J. (1999). Real-time water treatment process control with artificial neural networks. *Journal of Environmental Engineering*, *125*(2), 153–160. https://doi.org/10.1061/(ASCE)0733-9372(1999)125:2(153)

Zhu, J.-J., Segovia, J., & Anderson, P. R. (2015). Defining influent scenarios: Application of cluster analysis to a water reclamation plant. *Journal of Environmental Engineering,* *141*(7), 04015005. https://doi.org/10.1061/(ASCE)EE.1943-7870.0000934

Zounemat-Kermani, M., & Teshnehlab, M. (2008). Using adaptive neuro-fuzzy inference system for hydrological time series prediction. *Applied Soft Computing, 8*(2), 928–936. https://doi.org/10.1016/j.asoc.2007.07.011

6 Treatment of Wastewater by Application of Magnetic Field

Naval Koralkar
Department of Chemical Engineering, ITM (SLS) Baroda
University, Vadodara, Gujarat, India

CONTENTS

6.1 INTRODUCTION

In industries and domestic wastewater systems considerable technical issues and financial losses results due to scale deposition caused by obstructing water flow in pipes, lowering desalination efficiency, and reducing thermal transfer (1–6). $CaCO_3$, $CaSO_4$, $SrSO_4$, $BaSO_4$, CaF_2, $Ca3(PO_4)_2$, silica, and silicates are common scales found in water systems. Scaling happens when a salt's concentration is greater than its ability to dissolve in water. The most frequent causes include a concentration or evaporation process, changes in pH, temperature, outgassing, or pressure that impact the solubility of the salts (7, 8). When the temperature of the water rises, the solubility of $CaCO_3$ drops, resulting in precipitation onto heated surfaces.

DOI: 10.1201/9781003325147-6

Scale formation mainly depends on the quality of feed water. This includes the chemistry of the water. Alkaline (e.g., $CaCO_3$), non-alkaline (e.g., $CaSO_4$), and silica-based scales are the most common forms of scale (9, 10). The most prevalent component of scale is calcium and bicarbonate ions, which can be found in surface water, groundwater, brine, and water from industries (11, 12). Other salts that cause scale formation include $MgCO_3$, $BaSO_4$, $Fe_2(CO_3)_3$, iron oxides, silicates, fluorides, and phosphates (7, 13), and compounds with limited water solubility. Calcite, aragonite, and vaterite (14, 15) are three distinct crystal forms of the common hardness, $CaCO_3$. Calcite produces tougher scales, whereas aragonite and vaterite produce softer, easier to remove scales (16–19). In groundwater and wastewater, $CaSO_4$ and $Ca_3(PO_4)_2$ are typical scale elements. Al, Fe, Mg, and Ca (20, 21) are the most common metal hydroxide forms of amorphous silicic acid $[Si(OH)_4]$ found in silica and silicates. It is challenging to remove the sticky silica coating created by the supersaturation and polymerization of soluble silica (11).

Supersaturation, nucleation, crystallization, and precipitation are all involved in the development of scales (7, 22, 23). Supersaturation alone is insufficient for a solution to crystallize; instead, crystals form after supersaturation and nucleation and then grow out of solution (7). In order for crystals to form, particles, nuclei, or seeds must be present in a solution to serve as crystallization sites. Other small solubility salts, such as barium and strontium, frequently co-precipitate with $CaCO_3$, even though they have not yet reached saturation. Nucleation can be triggered in a variety of ways, including agitation and seeding (24, 25, 9, 26, 27). Surface/heterogeneous crystallization results from the lateral expansion of scale seeds trapped in openings, such as the walls of the vessel holding the solution (7). In saturated solutions, crystal seeds form in the bulk phase, leading to bulk/homogeneous crystallization.

Water-contact surfaces, such as heat exchangers, water pipes, and membranes, are referred to as foreign bodies in water systems. In all foreign substances, bulk and surface crystallization can happen separately or simultaneously. These systems have suffered significant economic losses as a result of the establishment of scale. Costs of membrane scaling include direct and indirect costs for routine maintenance, preparing feed water, using more energy to remove membrane scale, as well as indirect costs for reduced water output and limited membrane life (28).

To combat scaling, different chemical or physical therapies have been tried. Traditional methods for preventing the development of scale include ion exchange, the pre-precipitation of salts that are sparingly soluble, the use of chemicals, and the use of scale inhibitors. These procedures are costly, and they may alter the chemistry of the solution, posing health risks to humans and aquatic life (7). Additionally, the majority of the scale inhibitors are phosphate compounds, which could be harmful to the environment and result in eutrophication and algal blooms.

Examples of non-chemical water treatment methods include electromagnetic field (EMF), ultrasonic, catalytic materials, and alloys (11, 29–31). Ultrasonic waves have substantial mechanical and thermal impacts, as well as the ability to generate intense shock waves and microstreaming to avoid scaling (32). A little amount of catalytic minerals, like Zn, can reduce the pace at which calcium carbonate crystallizes and promote the crystallization of aragonite rather than calcite (30). For more than a century,

EMF has been employed as a scale control tool (33–35). Poter (36) was the first to suggest using electromagnetic fields (EMFs) as a non-chemical water treatment solution for controlling scale. Faunce and Cabell (37) developed an electromagnetic system to purify boiler feed water. In 1873, Hay became the first person to obtain a US patent for an EMF water treatment device (38). In addition to exploring potential EMF mechanisms, Baker and Judd (8) looked at industrial applications of commercial EMF devices from the preceding century. Salman et al. (39) emphasized both effective and ineffective anti-scaling effects of EMF studies, but he did not offer a comprehensive study of them. Ambashta and Sillanpaa (40) examined several aspects of magnetism and magnetic materials for water purification as well as a method for magnetically assisting water purification. With an emphasis on pulsed-EMF therapy, Piyadasa et al. (11) gathered the research on scaling and biofouling in reverse osmosis (RO) membranes and heat exchanger systems. Alabi et al. (7) published research on the stated impacts of EMF water treatment as well as potential mechanisms.

Although EMF has apparently proven useful for various industrial applications (41), its anti-scaling performance in water systems has a lengthy and contentious history. Previous work related to wastewater treatments have demonstrated that EMF is an important tool for controlling the scaling effect. In these publications, EMF has been utilized to improve oil separation and water splitting, decrease bacterial contamination, organic and inorganic fouling, and support other water treatment technologies like electrocoagulation and advanced oxidation processes. However, the multiple consequences and processes are unclear, and it has not yet been demonstrated empirically that exposure to EMFs is strong enough to cause significant anti-scaling effects. The inconsistent results of EMF investigations may be caused by the use of non-standardized procedures, variations in the water composition, or changes in the manner of operations (11). The composition of the pipe materials may have an impact on the effectiveness of magnetic water treatment. Standard operating procedures are usually ambiguous, and important details like the materials of the pipe, the duration of the exposure, and the field properties are only partially recorded. Regarding how these modifications will affect how EMF is used, there is little general consensus.

The present chapter examines the mechanics of EMF and the elements impacting the effectiveness of technologies. The influence of operational conditions on anti-scaling efficacy in water systems have received considerable attention. This chapter refers to electromagnetic field (EMF) therapies, which include electric field, magnetic field, and electromagnetic field. Water is passed through an EMF with certain qualities as part of the treatment. Pretreatment and cotreatment are terms used to describe the placement of EMF devices on treated water before it enters water systems or where a scaling surface is located.

6.2 MECHANISM OF MF ENHANCED WASTE WATER TREATMENT

To clarify the mechanisms of MF-enhanced WWT, two themes are offered: magnetic physico-chemistry and magnetic biology.

6.2.1   Physicochemical Effects of MFs on WWT

Since it interacts magnetically with electrons or atoms of pollutants in wastewater, MFs have considerable physicochemical impacts as a physical technique. Therefore, MFs not only increase the efficacy of removing contaminants from wastewater, but also provide a novel method for recycling magnetic adsorbents and catalysts throughout the WWT process. In this section, we discuss the water magnetization, the Lorentz force, and the magnetization effect.

6.2.1.1   Magnetization Effect

Eq. (6.1) can be used to express the magnetic field index (MFI) (*H*) (Maxwell equation). Eq. (6.2) can be used to calculate the relationship between the magnetic susceptibility (X_m) and magnetization (*M*), regardless of whether the materials are diamagnetic or paramagnetic (42).

$$H = \frac{1}{\mu_o} B - M \tag{6.1}$$

$$M = X_m H \tag{6.2}$$

where *H* is MF intensity, A.m^{-1}; 0 is vacuum permeability, $4\,4\pi \times 10^{-7}$ Wb.m^{-1}. A^{-1}; *B* is intensity of magnetic induction, *T*; *M* is magnetization, A.m^{-1}; X_m is a dimensionless quantity referred as magnetic susceptibility. Material magnetization in MFs can alter magnetoresistance (MR) and cause molecular rearrangement, both of which have a direct bearing on the effectiveness of WWT.

6.2.1.2   Magnetoresistance

The magnetoresistance effect, which was first proposed in the 1960s (43, 45), is characterized by the resistance variation of insoluble substances such as pollutants, catalysts, and adsorbents under the MFs.

$$R\% = \frac{R_H - R_0}{R_0} \tag{6.3}$$

where R_H and R_0 are the values of MR when the MFI are *H* and zero, respectively (45).

The impact of magnetic resonance imaging (MRI) might be favorable or negative. Stronger MR effects are produced by non-magnetic materials than by magnetic ones (43, 46). The interaction of lattice, spin, charge, and orbital degrees of freedom in a substance's electron systems is a major factor in the positive MR effect. The most significant parameters influencing this effect are the spin polarization characteristics of substances and the Lorentz force generated in MFs, which have been reported for temperature ranges between 70–340 K and MFI (zero to several Tesla) (45, 47, 48). The electron-hole recombination effect can be defined as a large number of electrons falling into holes under the influence of the Lorentz force, leading to electron-hole recombination. The Lorentz force produced by 0.05–1 T MFs can increase carrier collision probability. He et al. (43) thus achieved a 400 percent R and a good MR effect. The negative MR effect, which can be noticed in moderate MFs (46), refers to a reduction in insoluble material resistance in contrast to the positive MR impact. This effect is principally brought about by the suppression of electron spin fluctuations, which is connected to spin polarization and entails changing the spin polarity of electrons and nuclei (49, 50).

Adsorbents and catalysts are widely used in the WWT process. More charge carriers can engage in the surface reaction and form active materials like OH per unit time thanks to the MR impacts of these materials, especially the negative MR impact, which can induce charge carriers like Fe_2O_3 (Fe^{3+}) to relocate to active reaction sites. As a result, wastewater treatment is more effective. This conclusion was verified by Li and colleagues in a photocatalytic WWT process that was boosted by a 0.6 T SMF. The catalyst's ferromagnetic configuration facilitated oxygen evolution reaction by attracting spin-oriented electrons at the interface (51).

6.2.1.3   Rearrangement of Particle

Molecules are classified as polar or non-polar depending on whether their positive and negative charges coincide (42). Polar molecules have opposite positive and negative charge distributions, whereas non-polar molecules have opposite positive and negative charge distributions and have a zero-dipole moment. By modifying the electrokinetic potential of charged particles, MFs have the ability to alter molecule polarization, conductivity, and structure (52). MFs can eventually produce a shift in electrokinetic potential, which can result in charged particle rearrangement (53).

Paramagnetism and diamagnetism are two further classifications for materials. The conflict between electronic spin-exchange, spin orientation, and electron thermal mobility, which is mediated by magnetic interactions, is the main cause of the effects of MFs on these materials (54–56). Paramagnetic molecules have the ability to reorganize themselves in the MFs' direction. This rearrangement event is caused by the MFI, which in mild MFs is relatively weak (57). Diamagnetic states primarily mediate weak magnetic interactions, which have a low interaction energy compared to thermal energy (55). Due to their spontaneous magnetization, ferromagnetic molecules in MFs, on the other hand, can be easily changed.

$$\mu = \frac{V_m J_s}{N_A} \qquad (6.4)$$

The molar volume of a ferromagnetic atom is $m^3 \cdot mol^{-1}$; the saturated magnetic polarisation is $A \cdot m^{-1}$; and the Avogadro constant is N_A. With the foregoing in mind, the positive and negative charges of ferromagnetic molecules can be easily separated due to the difference in magnetic moment caused by MFs (Figure 6.1) (58, 59). Thus, despite their initial disordered arrangement in wastewater, ferromagnetic molecules can be easily reorganized (60).

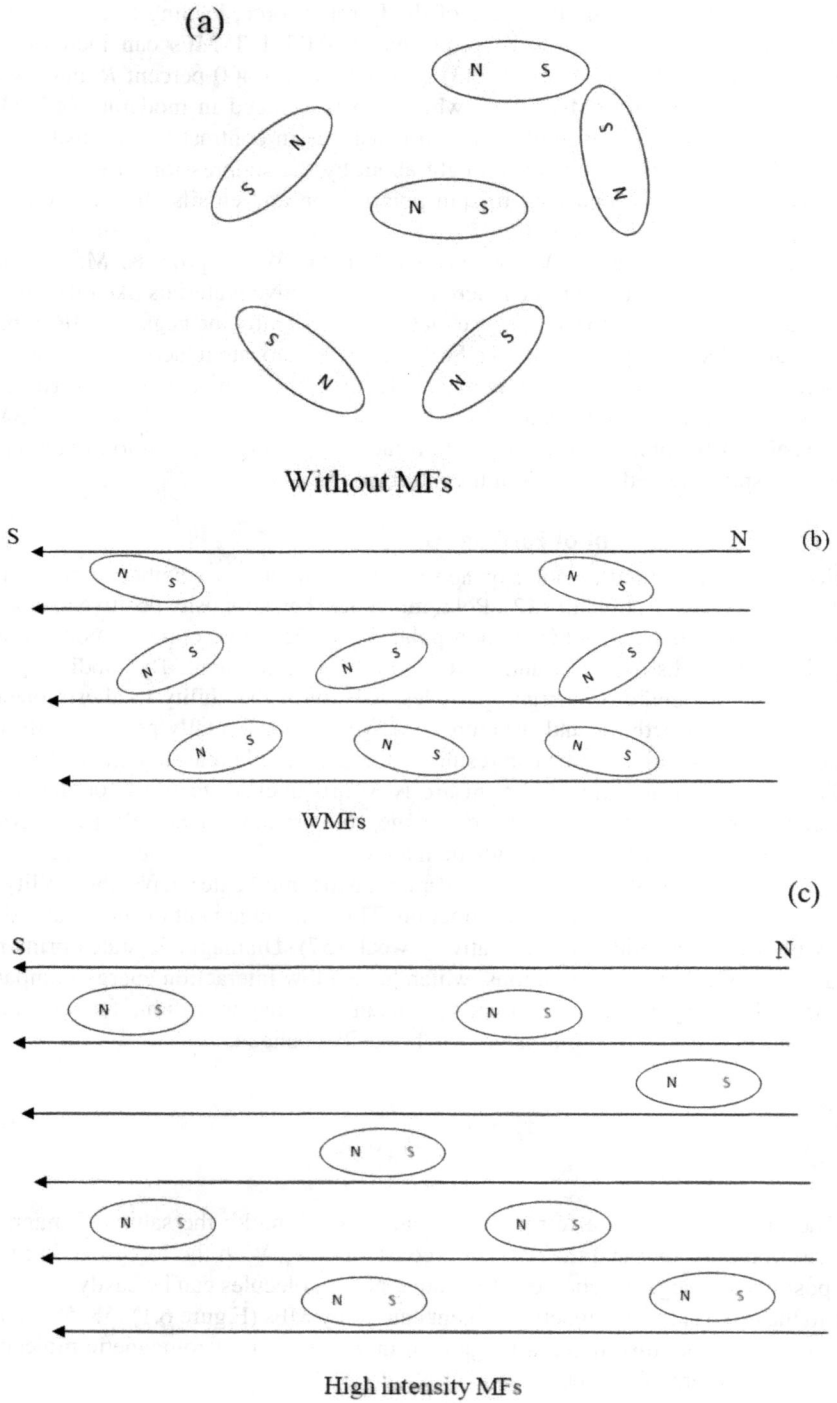

FIGURE 6.1 Particle rearrangement in MFs (63).

Omar et al. (59) found that 0.009–0.03 TMF (moderate SMF) was beneficial during collisions between particles and activated sludge in WWT by rearranging particles in accordance with magnetic induction lines. This charged particle rearrangement increased dramatically with the MFI, leading to more severe coagulation and congregation (Figure 6.1) (42). MFs can help with processes such as activated sludge sedimentation (61), granulation (62), and the removal of particles and contaminants from the WWT process.

6.2.2 LORENTZ FORCE

The force acting on ions traveling in MFs is known as the Lorentz force. Because of the Lorentz force's ability to reorient moving ions, the charged particle reaction interface's Zeta potential is altered (64, 65).

6.2.2.1 Lorentz Force

Lorentz force is produced when charged particles interrupt magnetic induction lines of MFs at a velocity of v (Eq. 6.5) (66). Ions are forced by the Lorentz force to move in moderate MFs (0.01–0.1 T) from the high-speed fluid layer to the low-speed fluid layer, creating "ion condensation" close to solid interfaces like catalyst surfaces. The MFI increases as the Lorentz force's influence grows (67). Eq. (6.6) can be used to express the fluid's flow rate (Hagen–Poiseuille equation).

$$F_L = qvB \tag{6.5}$$

$$v_z = v_0 z (h - z) \tag{6.6}$$

where v_0 denotes the maximum velocity (v_{max}) at half-height relative to the solid interface; F_L denotes the Lorentz force in units of N; v denotes the velocity of charged particles in units of m/s; q denotes the charge of charged particles in units of C; and v_z equals 0 at both the top and bottom ($z = h$ and 0). Gravity and magnetic forces acting on ions are, respectively, $1–10^2$ times and $10^3–10^4$ times weaker than the Lorentz forces generated by a 0.3 T SMF. Eq. (6.7) can be used to compute the magnetic force (66).

$$F_M = \left(\frac{X_{mol} B}{\mu_o} \frac{\partial B}{\partial i} \right)_{i = x, y, z} \tag{6.7}$$

The other parameters are the same as in Eqs. (6.1) and (6.2). Here $\frac{\partial B}{\partial i}$ is the MF gradient in the I direction; F_M is the magnetic force that acts on ions, Nmol^{-1}; the molar susceptibility is denoted by Xmol, mol^1; and the additional variables are the same as in Eqs. (6.6) and (6.7). As a result, the Lorentz force has an impact on the ions' original track and direction as they pass through MFs (Figure 6.2) (68).

(a)

(b)

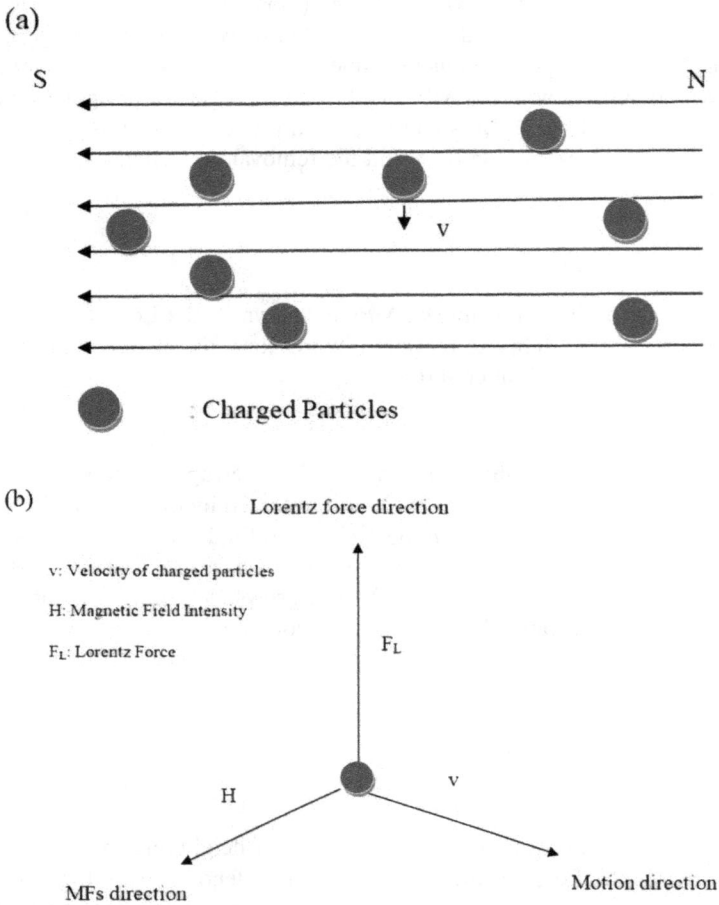

FIGURE 6.2 Charged particle movement and Lorentz force analysis in MFs. (a) Charged particle movement in MFs; (b) charged particle Lorentz force analysis in MFs (63).

The mechanism of MF-enhanced WWT, such as how altering the collision probability and changing the direction of charged particles on the reactive surface can increase WWT effectiveness, has frequently been described using the Lorentz force (68, 69).

According to Ma et al. (70), the Lorentz force can boost the adsorption of methylene blue by reorienting dye molecules when SMF is present (0.2 T). The Lorentz force acting on dye molecules in a rotating magnetic field (0.2 T, 2400 r/min) was multidirectional, greatly increasing the possibility of dye molecules crashing onto adsorbents. Furthermore, the removal of ferromagnetic heavy metals like Cr (IV) can be aided by the Lorenz force in an SMF (0–0.002 T) (71).

Furthermore, Pirsaheb et al. (72) discovered that under the influence of weak MF, the Lorentz force can enhance photocatalytic performance by promoting quick mass transfer and ion condensation in reaction sites.

6.2.2.2 Zeta Potential

An electric double layer is composed of the particle surface, stern layer, and diffusion layer. Zeta potential is the potential in an electric double layer between a specific location in the sliding plane and fluid far from the interface (64). It has been demonstrated that the Lorentz force acting on charged particles in a 2 T MF causes ion movement deflection, which causes charge instability and a drop in Zeta potential from 40 mV to 16.5 mV. This effect, known as the Lorentz ion shift (X_i), is more pronounced close to the particle surface (64). The following is a representation of ΔX_i:

$$\Delta X_i = \frac{q}{6\pi\eta r_i} Btv \qquad (6.8)$$

where r_i is the ion radius, in m, t is the particle exposure time in MF, in s, X_i is the Lorentz ion shift, in m, and ΔX_i is the fluid viscosity, in kgs^{-1} m^{-1}. The other requirements are identical to those mentioned above. The stability of a colloidal system is frequently assessed using the Zeta potential. Suspended solid stability and settling characteristics are easily influenced by the Zeta potential (73, 74).

Here ΔX_i is the Lorentz ion shift, in m; is the fluid viscosity, in kgs^{-1} m^{-1}; t is the particle exposure time in MF, in s; and r_i is the ion radius, in m. The other criteria are the same as the ones listed above. The Zeta potential is commonly used to determine colloidal system stability. The Zeta potential can readily impact the stability and settling characteristics of suspended solids (73, 74). The increase in ion-exchange (Eq. (6.9) between pollutants and adsorbents was discovered to be the main impact of a 0.57 T SMF in WWT process on Zeta potential (75).

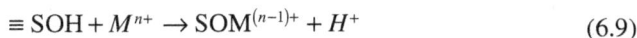

$$\equiv SOH + M^{n+} \rightarrow SOM^{(n-1)+} + H^+ \qquad (6.9)$$

where M^{n+} is a cation in wastewater.

6.2.2.3 Water Magnetization

The most prevalent solvent, water, is magnetically sensitive (76). MFs can have a big impact on water parameters such self-diffusion coefficient (77), evaporation rate (78), and water cluster size (79). Changes in molecular structure, polarization, and particle arrangement, as well as changes in electro-kinetic potential generated by MFs, all contribute to changes in water properties (53). The rapid coagulation of particles in water can be triggered by the electro-kinetic potential lowering of water generated by MFs (53, 80). In 0.4–0.6 T SMF, Krzemieniewski et al. (53) found that the electro-kinetic potential of water decreased, the precipitation time was shortened, and the coagulation of the sludge particles was improved. Furthermore, Hakobyan and Ayrapetyan (81) found that the conductivity of deionized water can be decreased by a 0.0025 T SMF, which they attribute to a change in the ionic hydration shell (82).

(a)   (b)

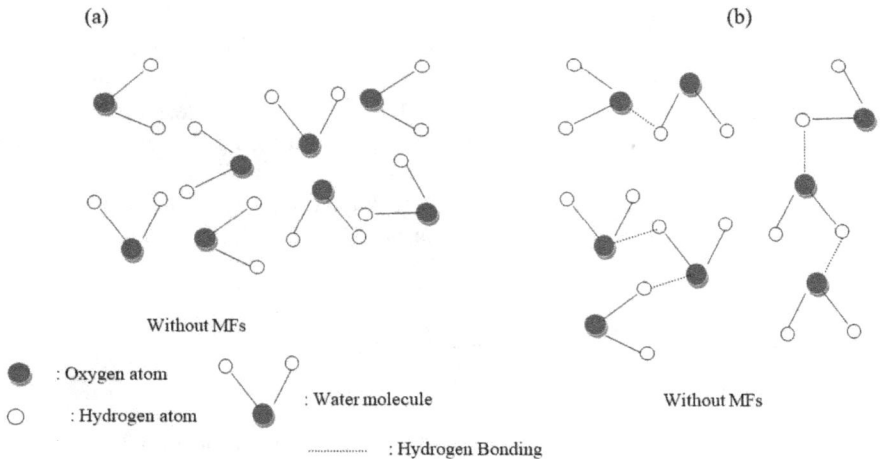

Without MFs

● : Oxygen atom

○ : Hydrogen atom

: Water molecule   Without MFs

: Hydrogen Bonding

FIGURE 6.3   Water molecules hydrogen bonds are strengthened by MFs (a) hydrogen bond of the flexible three-centered (F3C) water model without MF proposed by Levitt et al. (85), and there are three interaction centres and (b) water molecule's hydrogen bond with MF (Wang et al. (63)).

   The three-dimensional hydrogen bonding network is hypothesized to be connected to some water properties, including structural stability and viscosity (76, 83, 84). Chang and Weng (79) found that SMF (0–10 T) can increase the amount of hydrogen bonds (0–0.34 percent) using the flexible three centred (F3C) water model (Figure 6.3). Viscosity and water stability rose, despite the lower self-diffusion coefficient. Inaba et al. on examining the 6 T MF-enhanced H_2O and D_2O (heavy water) transition found that the hydrogen bonds were strengthened as a result of the Lorentz force preventing the thermal motion of charged atoms, a phenomenon known as "enhanced-dynamic magnetic susceptibility". Furthermore, the shorter distance and stronger bonding force of hydrogen bonds may be achieved because of the somewhat organized arrangement of water molecules in MFs (77). Moderate MFs (1 T) have been reported by Cai et al. (87) to encourage the formation of hydrogen bonds between water molecules in addition to strong and ultra-strong MFs. On the other hand, Toledo et al. (88) proposed that MFs increase the hydrogen link inside clusters while weakening the hydrogen bond between water clusters. According to Wang et al. (89), the friction coefficient was lower in magnetized water than in unmagnetized water, and this friction difference might be attributable to the MF's diminished impact on hydrogen bonding between water molecules. In general, MF effects occur in altered hydrogen bonds between water molecules, which increases the contributions of water to reactions like adsorption enhancement (90).

6.3   MF'S BIOLOGICAL EFFECTS IN WWT

This section delves further into the intracellular radical pair process, enzyme activity, and cell membrane. It is important to note that there is not necessarily a linear relationship between these variables and MF effects, as the "biological window effect" illustrates (91, 92, 93).

6.3.1 CELL MEMBRANE

By altering the cell's shape and surface hydrophobicity, the membrane's phospholipid fatty acid can be made to face in a different direction (96, 97). MFs impact membrane fluidity and cell metabolism in addition to cell membrane potentials (98). By smoothing the cell membrane and raising its permeability and fluidity at low temperatures, Niu et al. (91) found that a MF of 0.013 T boosted fatty acid synthesis in the cell membrane. The flocculation ability of activated sludge in the WWT system was aided by the increased relative hydrophobicity and decreased membrane negative charge caused by MFs, causing the greatest reduction in chemical oxygen demand (99). The impact of MFs on the integrity of cell membranes, on the other hand, cannot be overlooked. Although self-repair of the membrane strives to mitigate this influence, carbohydrate and energy metabolism are inevitably altered (100).

The ion channel in the cell membrane can also be excited by the Lorentz force acting on the ions on the membrane (101, 102). Numerous ions (Na^+, K^+, Ca^{2+}, etc.) are found on both sides of the cell membrane, and their concentrations influence the transmembrane ion channel switch and cell volume. Additionally, the ion channel is crucial for the transmission of signals (103). In 0.06 T MF, the Na^+/K^+ ion channel activity was found to be increased, whereas in 0.125 T MF, it was shown to be inhibited (104, 105). Lin et al. (106) and Wang et al. (107) discovered that the MF-induced change in ion signal channels has a variety of biological impacts, including improved cell survival and altered cellular structure.

6.3.2 ACTIVITY OF ENZYME

Enzymatic systems control the effectiveness of biological WWT as catalyzts and "primary magneto sensors" for all biochemical reactions. They are magneto sensitive to MFs (108, 109).

In general, MFs can either increase or decrease enzyme activity (95), such as tryptophane deaminase activity in a 0.04 T MF or dehydrogenase activity in a 0.007 T MF (90, 110). Further investigation demonstrated that low frequency MF facilitated glucose uptake by increasing the activity of glutamate dehydrogenase and glucose dehydrogenase (111). Laccase activity was dramatically boosted in fungus after exposure to 10–40 Hz rotating MF (112). By boosting the concentration of exoenzymes in activated sludge in WWT, MFs can also enhance the biodegradation capacity of the material (113).

MFs can alter activity of enzyme by (1) modifying the molecular structure of enzymes by suppressing stretching vibration in some chemical bonds, including secondary and tertiary structure of pertinent proteins, for example, when subjected to a SMF of 0.2–1.34 T and a 50 Hz electromagnetic field with MFI of 0.0018 T (114, 115); (2) modifying the kinetic energy of unpaired electrons in the enzymes' catalytic activity center (112); (3) metal ions acting as coenzymes underwent changes in their electronic energy level transition, which affected the rate at which enzymes catalyzed reactions (109, 116, 117); (4) affecting electron transport and creating an enzyme structure that is more active in processes that use enzymes (115, 118).

6.4 APPLICATION OF MAGNETIC FIELD FOR WASTE WATER TREATMENT

Depending on whether the magnetic field intensity (MFI) changes over time, MFs can be classed as either dynamic (DMF) or static (SMF) (119).

6.4.1 EFFECT OF MAGNETIC FIELD ON HEAVY METALS PRESENT IN WASTEWATER

Depending on whether the magnetic field intensity (MFI) changes over time, MFs can be classed as either dynamic (DMF) or static (SMF) (75). The SMF can improve heavy metal adsorption-separation efficiency in wastewater. The employment of magnetic nanoparticles and MFs has generated far greater interest than the use of SMF alone due to the combination's simplicity of separation, large surface area, and changed surface electronegativity (120, 121).

To maximize absorbent recovery, the gas-assisted low-field magnetic separation (0.2 TMF) was created. The highest recovery efficiency of the Pb^{2+} adsorbing adsorbent Fe_3O_4/humic acid sodium was 85% (122).

Szatylowicz and Skoczko (123) indicated that pertinent research of MF-enhanced heavy metals' removal methods might show that MF could change the structure of suspended solids by producing Helmholtz free energy (F, J).

$$dF = -SdT - \mu dB \qquad (6.10)$$

where S is the system's entropy $(J.K^{-1})$, T is the absolute temperature (K), and μ is the system's total magnetic moments $(J.T^{-1})$, and B is the magnetic induction intensity (T). If B is high enough, particles in MFs are exposed to F, resulting in changes in particle energy structure and characteristics on a macroscale, notably for adsorbents (124). Tireli et al. (125) found that the maximum adsorption capacity of adsorbents and the Langmuir parameters both increased significantly in the presence of 5×10^{-5} T MF.

6.4.2 EFFECTS OF MFs ON THE TREATMENT OF DYE WASTEWATER

The environment and public health are at risk due to the high chroma and toxicity of the dyestuff discharge from the textile printing and dyeing industry. MFs had been shown to facilitate the adsorption degradation of dyes (68). Tireli et al. (125) found that employing both pillared and magnetic clay as adsorbents under 5×10^5 T SMF significantly increased the adsorption rate of methylene blue. The combination of zerovalent iron/Cu^{2+}/peroxymonosulfate combined with a modest magnetic field increased the adsorption efficiency of acid Orange 7 from 60% to 95% as the corrosion of zerovalent iron was accelerated. During this reaction, Cu^{2+} was reduced to one Cu^+

during reaction, and other Cu^+ was the main ion activating the peroxymonosulfate to create free radicals (OH and SO_4) (126).

Improved dye adsorption can result from the combination of magnetic nanoparticles with MFs. A rotating MF of 0.01899 T and a polypyrrole magnetic nanocomposite may be used to achieve the highest rate of Congo red elimination by increasing the magnetic force and redirecting the charged particles (127). Magnetic fields and magnetic nanoparticles can function alone as well. Magnetic nanoparticles increase the adsorption rate, whereas MFs are only needed to recover the magnetic nanoparticles. Ag_3PO_4-TiO_2-Fe_2O_3, which can be quickly separated with external MF, was created by combining Ag_3PO_4-TiO_2 and magnetic Fe_2O_3 powder, according to An et al. (128). These combinations produced a special color removal technology with high adsorption and recycling capacities (129).

According to Tan et al. (130), Song et al. (131), and Wang et al. (132) MFs may accelerate dye biodegradation by enhancing microbial metabolism and enzyme activity. They also discovered that dyes could be efficiently biodegraded. Pichia occidentalis A2 demonstrated improved dye degradation efficiency and salt tolerance at a SMF of 0.2063 T, which can be attributed to elevated manganese peroxidase and laccase activity (132). Similar to this, a SMF of 0.095 T facilitated Acid Red B biodegradation in a sequencing batch reactor, and Acid Red B removal efficiency in a high salt environment was over 98% (130). Improvements were made to the system's lignin peroxidase and laccase activity, as well as its biomass content and sludge settleability (133). Additionally, spinning the MF at 1 to 50 Hz and 0.005 to 0.018 T can greatly increase the activity of horseradish peroxidase over a wide pH range. Consequently, it is possible to significantly increase the effectiveness of biocatalytic dye degradation (134).

Both pollutants and adsorbents can be affected by MFs due to their physicochemical features (68, 135). According to Ma et al. (136), a rotating MF with an intensity and frequency of 0.2 T and 2400 r/min, respectively, changed the Zeta potential of the adsorbent surface. As a result, electrostatic and interaction-based adsorption behaviors have changed (90). Tireli et al. (125) found that a 5×10^{-5} T magnetic field might enhance the adsorption capacity of methyl blue by altering the surface heterogeneity of biochar and increasing the specific surface area. Reorienting dye molecules enhances the likelihood of colliding with adsorbents under the influence of Lorentz force in MFs of 0.01–0.4 T, which can boost adsorption efficiency (68, 69, 127). In general, the treatment of dye wastewater with magnetic adsorption separation is effective.

6.4.3 Effects of MFs on the Treatment of Domestic Wastewater

Primary, secondary, and tertiary treatment are common home wastewater treatment processes, with secondary treatment typically being a biological process (137). During first treatment, MFs can significantly boost the removal of turbidity and suspended particles from wastewater (138). The sedimentation tank's footprint could be lowered as treatment efficiency improves, which is important for urban wastewater treatment plants (WWTPs) (139).

MFs' influence on biological household wastewater treatment procedures, such as pollutant removal (61), biofilm formation (140), and enzyme activity (42) have been investigated. Xu et al. (141) discovered that a SMF of 0.03 T improved bacterial and gene abundances, and enzyme activity. As a result, the effectiveness of removing nitrogen from ammonia $(NH_4^+ - N)$ and total nitrogen (TN) rose by 22.4 percent and 39.5 percent, respectively. Lebkowska et al. (92) found that a 0.0075 T SMF increased p-nitroaniline breakdown, which reduced chemical oxygen demand (COD) by 25%.

6.4.4 Photomagnetic Coupling Technology

Due to its cost-savings, efficiency, and oxidizing potential, photocatalysis, like the photo-Fenton process, is an advanced oxidation process in WWT. MF is a successful strategy in WWT to increase photocatalytic degradation. According to Li et al. (142), a local MF produced by plasma Ag in Ag-BiVO$_4$-MnO$_x$ boosted the likelihood of electron-hole pair formation and photogenerated electron-hole pair migration. Ag-photocatalytic BiVO$_4$-MnO$_x$'s performance was enhanced as a result of the production of the dominant active specie O$_2$. When the MFI was 0.15 T, the degradation of Congo red enhanced by magnetized photocatalyst CoFe$_2$O$_4$/MoS$_2$ had the highest degradation efficiency (143). Using MFs significantly increased the efficiency of rhodamine B degradation, which rose by 3.2% as the MFI increased from 0.01 T to 0.08 T (144). Additionally, higher removal efficiencies of 94.5 percent and 97.4 percent for $\left(NO_3^- - N\right)$ and $\left(NH_4^+ - N\right)$ were attained in 0.3 T MF by applying Lorentz force to encourage molecules to approach the catalyst surface (145).

One significant drawback of photocatalysis in WWT is the inability to recycle photocatalysts (146). By covering photocatalysts with iron oxides (such as Fe$_3$O$_4$, Fe$_2$O$_3$) to make them ferromagnetic, it is possible to accomplish the heterogeneous photo Fenton (HPF) reaction and get around the WWT method's restriction on photocatalyst recycling (147). The breakdown of refractory organic compounds can be improved and separation efficiency can be increased by adding MF after coating Fe$_3$O$_4$ on the surface of TiO$_2$-graphene oxide (148). Diatomite-Fe$_2$O$_3$-TiO$_2$ composites have been demonstrated by Barbosa et al. (149) to be capable of degrading methyl blue in the dye WWT. The amount of free radicals significantly increased during the photodegradation process due to an electron transfer between Fe^{3+} and Fe^{2+} mediated by TiO$_2$ (e.g., OH, O$_2$, HO$_2$). Additionally, in a 1.5 T SMF, a Fe$_3$O$_4$/SnO$_2$ magnetic nanocomposite with good separation and recovery capabilities effectively eliminated crystal violet in wastewater (146). After being recycled by MFs five times, the magnetic photocatalyst with 6.6 percent Fe can still remove 92 percent of metronidazole (150).

The result of the radical A radical-induced reaction is the MF-enhanced photo-Fenton catalytic WWT (151, 152). Leaching of iron ions from the photocatalyst surface during the electron transition from Fe^{2+} to Fe^{3+} in solution can accelerate photo-Fenton degradation by generating free radicals OH (153). As a result, iron oxide-coated photocatalysts can be regenerated by MFs while simultaneously improving

OH generation by transferring electrons between Fe^{2+} and Fe^{3+} (Eqs. (6.11)–(6.12)) (150). Barbosa et al. (149) similarly obtained similar results (Eqs. (6.12)–(6.13)). When the degradation of methyl orange was accelerated by photo-magnetic coupling technology, the Lorentz force and "ion condensation" that occurred at the reaction interface of the photocatalyst affected the movement channel of OH (Eq. (6.14)) (67). Additionally, a concentration of spin-oriented electrons accumulated at the reaction interface as a result of the photocatalysts' ferromagnetic alignment, which may increase the activity of the oxygen evolution event. (154).

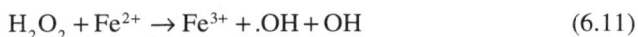

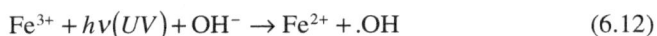

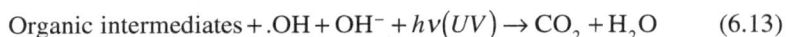

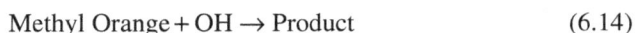

$$H_2O_2 + Fe^{2+} \rightarrow Fe^{3+} + .OH + OH \tag{6.11}$$

$$Fe^{3+} + h\nu(UV) + OH^- \rightarrow Fe^{2+} + .OH \tag{6.12}$$

$$\text{Organic intermediates} + .OH + OH^- + h\nu(UV) \rightarrow CO_2 + H_2O \tag{6.13}$$

$$\text{Methyl Orange} + OH \rightarrow \text{Product} \tag{6.14}$$

6.5 APPLICATION OF MFS IN WWT: CHALLENGES AND OPPORTUNITIES

6.5.1 CHALLENGES

1. There are two obstacles to the bio-WWT fostered by MFs:
   a. Due to the different MF operating factors, such as intensity, frequency, and action form, which may have either positive or negative magnetic physicochemical and biologic effects, full digital control of MF parameters is challenging in actual WWTP (155, 156, 157).
   b. Given the size, construction style, and MFs shielding of a WWTP, using a MF generator there is hard.
2. Magnetic contaminants can be eliminated directly via magnetic adsorption-separation (124). Non-magnetic contaminants, on the other hand, can only be separated via direct magnetic adsorption (158). Although magnetic adsorbents can overcome this restriction (150), further work is required to increase their adsorption efficiency, recovery efficiency, and the impact of recovery frequency on adsorption efficiency before they can be applied in the industrial setting.

6.5.2 OPPORTUNITIES

Nevertheless, MFs are particularly intriguing due to their high efficiency and economy, despite the complexity and difficulty of using them.

There are still many choices for WWTP startup and optimization since MFs can be used for decades with a single investment (42).

(1) The start-up and treatment effectiveness of the WWT process can be accelerated with the aid of MFs (94). It is important to look at the proper operating settings of MFs for the startup phase and WWT process optimization.

(2) MFs can increase the activity of the electron transfer chain in microbial electron transfer systems by changing the concentration and activity of electron donors like NADH/NAD+ (96, 159). The use of MFs thus creates new opportunities for studying the basic metabolic processes that underlie diverse bio-WWT processes.

(3) The future of WWT's novel strategy, MF-assisted adsorption, is bright (160). Many magnetic adsorbents, in particular magnetic nanoparticles coated with superparamagnetic iron oxide like Fe_3O_4 and Fe_2O_3, have lately been employed in heavy metal removal, dye adsorption, and photocatalysis due to their huge specific surface area, high reactivity, and cyclability (120, 121, 158). Emphasis should be placed on the combination of paramagnetic materials with conventional adsorbents, as well as the more affordable sources and efficient recovery method of magnetic adsorbents supported by MFs, for better MF-assisted adsorption.

6.6 SUMMARY

Despite several drawbacks, MF is still a solid method for enhancing WWT. In comparison to the conventional WWT process, the MF-enhanced WWT method has demonstrated favorable effects on adsorption separation, bio-WWT, and advanced oxidation process. The usage of MFs also offers enormous and exciting possibilities for WWT growth because to its environmental friendliness and high efficiency. The identification of mechanisms of MF-enhanced WWT is necessary for expanding the use of MF in it as well as for overcoming its drawbacks and benefiting from it.

REFERENCES

1. Gabrielli, C., Jaouhari, R., Maurin, G. & Keddam, M. Magnetic water treatment for scale prevention. Water Res. 35, 3249–3259 (2001).
2. Colic, M., Chien, A. & Morse, D. Synergistic application of chemical and electromagnetic water treatment in corrosion and scale prevention. Croat. Chem. Acta 71, 905–916 (1998).
3. Xu, P. et al. Critical review of desalination concentrate management, treatment and beneficial use. Environ. Eng. Sci. 30, 502–514 (2013).
4. Xu, X. et al. Use of drinking water treatment solids for arsenate removal from desalination concentrate. J. Colloid Interf. Sci. 445, 252–261 (2015).
5. Lin, L., Xu, X., Papelis, C. & Xu, P. Innovative use of drinking water treatment solids for heavy metals removal from desalination concentrate: Synergistic effect of salts and natural organic matter. Chem. Eng. Res. Des. 120, 231–239 (2017).
6. Lin, L., Xu, X., Papelis, C., Cath, T. Y. & Xu, P. Sorption of metals and metalloids from reverse osmosis concentrate on drinking water treatment solids. Sep. Purif. Technol. 134, 37–45 (2014).
7. Alabi, A., Chiesa, M., Garlisi, C. & Palmisano, G. Advances in anti-scale magnetic water treatment. Environ. Sci.: Water Res. Technol. 1, 408–425 (2015).

8. Baker, J. S. & Judd, S. J. Magnetic amelioration of scale formation. Water Res. 30, 247–260 (1996).
9. Antony, A. et al. Scale formation and control in high pressure membrane water treatment systems: A review. J. Membr. Sci. 383, 1–16 (2011).
10. Patel, S. & Finan, M. A. New antifoulants for deposit control in MSF and MED plants. Desalination 124, 63–74 (1999).
11. Piyadasa, C. et al. The application of electromagnetic fields to the control of the scaling and biofouling of reverse osmosis membranes-A review. Desalination 418, 19–34 (2017).
12. Colic, M. & Morse, D. Effects of amplitude of the radiofrequency electromagnetic radiation on aqueous suspensions and solutions. J. Colloid Interf. Sci. 200, 265–272 (1998).
13. Amjad, Z. Scale inhibition in desalination applications: an overview. Corrosion, NACE 96–230 (1996).
14. Plummer, L. N. & Busenberg, E. The solubilities of calcite, aragonite and vaterite in CO_2-H_2O solutions between 0 and 90 C, and an evaluation of the aqueous model for the system $CaCO_3$-CO_2-H_2O. Geochim. Cosmochim. Acta 46, 1011–1040 (1982).
15. de Leeuw, N. H. & Parker, S. C. Surface structure and morphology of calcium carbonate polymorphs calcite, aragonite, and vaterite: an atomistic approach. J. Phys. Chem. B 102, 2914–2922 (1998).
16. Xing, X., Ma, C. & Chen, Y. Investigation on the electromagnetic anti-fouling technology for scale prevention. Chem. Eng. Technol. 28, 1540–1545 (2005).
17. Kobe, S., Dražić, G., McGuiness, P. J. & Stražišar, J. The influence of the magnetic field on the crystallisation form of calcium carbonate and the testing of a magnetic water-treatment device. J. Magn. Magn. Mater. 236, 71–76 (2001).
18. Knez, S. & Pohar, C. The magnetic field influence on the polymorph composition of $CaCO_3$ precipitated from carbonized aqueous solutions. J. Colloid Interf. Sci. 281, 377–388 (2005).
19. Coey, J. & Cass, S. Magnetic water treatment. J. Magn. Magn. Mater. 209, 71–74 (2000).
20. Hater, W. et al. Silica scaling on reverse osmosis membranes-Investigation and new test methods. Desalin. Water Treat. 31, 326–330 (2011).
21. Bremere, I. et al. Prevention of silica scale in membrane systems: Removal of monomer and polymer silica. Desalination 132, 89–100 (2000).
22. Demopoulos, G. Aqueous precipitation and crystallization for the production of particulate solids with desired properties. Hydrometallurgy 96, 199–214 (2009).
23. Chen, T., Neville, A. & Yuan, M. Calcium carbonate scale formation-Assessing the initial stages of precipitation and deposition. J. Petrol. Sci. Eng. 46, 185–194 (2005).
24. Mullin, J. Crystallization, 4th Edition, Butterworth Heinemann (London, UK Oxford, 2001).
25. Youngquist, G. R. & Randolph, A. D. Secondary nucleation in a class II system: Ammonium sulfate-water. AIChE J. 18, 421–429 (1972).
26. Lee, S. & Lee, C.-H. Effect of operating conditions on $CaSO_4$ scale formation mechanism in nanofiltration for water softening. Water Res. 34, 3854–3866 (2000).
27. Lee, S., Kim, J. & Lee, C.-H. Analysis of $CaSO_4$ scale formation mechanism in various nanofiltration modules. J. Membr. Sci. 163, 63–74 (1999).
28. Avlonitis, S., Kouroumbas, K. & Vlachakis, N. Energy consumption and membrane replacement cost for seawater RO desalination plants. Desalination 157, 151–158 (2003).

29. Broekman, S., Pohlmann, O., Beardwood, E. & de Meulenaer, E. C. Ultrasonic treatment for microbiological control of water systems. Ultrason. Sonochem. 17, 1041–1048 (2010).
30. Coetzee, P., Yacoby, M., Howell, S. & Mubenga, S. Scale reduction and scale modification effects induced by Zn and other metal species in physical water treatment. Water SA 24, 77–84 (1998).
31. Tijing, L. D. et al. Mitigation of scaling in heat exchangers by physical water treatment using zinc and tourmaline. Appl. Therm. Eng. 31, 2025–2031 (2011).
32. Hou, D., Zhang, L., Fan, H., Wang, J. & Huang, H. Ultrasonic irradiation control of silica fouling during membrane distillation process. Desalination 386, 48–57 (2016).
33. Lipus, L. C., Ačko, B. & Hamler, A. Electromagnets for high-flow water processing. Chem. Eng. Process. 50, 952–958 (2011).
34. Vallée, P., Lafait, J., Mentré, P., Monod, M.-O. & Thomas, Y. Effects of pulsed low frequency electromagnetic fields on water using photoluminescence spectroscopy: role of bubble/water interface. J. Chem. Phys. 122, 114513 (2005).
35. Koza, J. A. et al. Hydrogen evolution under the influence of a magnetic field. Electrochim. Acta 56, 2665–2675 (2011).
36. Porter, A. F. Preventing incrustation of steam boilers. U.S. Patent 50,774 (1865).
37. A. Faunce, S. C. Electric means for preventing boiler incrustation. U.S. Patent 438,579 (1890).
38. Hay, A. T. Electrical protection for boilers. U.S. Patent 140,196 (1873).
39. Salman, M., Safar, M. & Al-Nuwaibit, G. The effect of magnetic treatment on retarding scaling deposition. TOJSAT 5, 62–67 (2015).
40. Ambashta, R. D. & Sillanpaa, M. Water purification using magnetic assistance: A review. J. Hazard. Mater. 180, 38–49 (2010).
41. Baker, J. S., Judd, S. J. & Parsons, S. A. Antiscale magnetic pretreatment of reverse osmosis feedwater. Desalination 110, 151–165 (1997).
42. Zaidi, N. S., Sohaili, J., Muda, K., Sillanpaa, M. Magnetic field application and its potential in water and wastewater treatment systems. Sep. Purif. Rev. 43, 206–240 (2014).
43. He, X., Yang, Z., Zhu, C., He, B., Luo, F., Wei, P., Zhao, W. Y., Wang, J. F., Sun, Z. G. Negative differential resistance and unsaturated magnetoresistance effects based on avalanche breakdown. J. Phys. 32, 8 (2020).
44. Li, J., Qi, P., Wang, Y. R., Zhou, Y., Zhang, Z. M., Cao, Q. Q., Wang, D. H., Mi, W. B., Du, Y. W. Enhanced photocatalytic performance through magnetic field boosting carrier transport. ACS Nano 12, 3351–3359 (2018b).
45. Nguyen, L. H., Hung, L. X., Phuc, N. X., Nam, P. H., Ngan, L. T. T., Dang, N. V., Bau, L. V., Linh, P. H., Phong, P. T. Composites (1x) $La_{0.7}Ca_{0.3}MnO^3$/ $xLa_{0.7}Sr_{0.2}Ca_{0.1}MnO_3$: Electrical transport properties and enhancing of low-field-magnetoresistance and colossal magnetoresistance. J. Alloy. Compd, 156607 (2020).
46. Mitani, Y., Fuseya, Y. Large longitudinal magnetoresistance of multivalley systems. J. Phys.-Condes. Matter. 32, 7 (2020).
47. Li, X. W., Gupta, A., Xiao, G., Gong, G. Q. Low-field magnetoresistive properties of polycrystalline and epitaxial perovskite manganite film. Appl. Phys. Lett. 71, 1124 (1997).
48. Phan, T. L., Zhang, P., Grinting, D., Yu, S. C., Nghia, N. X., Nguyen, V. D. Influences of annealing temperature on structural characterization and magnetic properties of Mn doped $BaTiO_3$ ceramics. J. Appl. Phys. 48–4018 (2012).

49. Mesquita, F., Magalhaes, S. G., Pureur, P., Diop, L. V. B., Isnard, O. Electrical magnetotransport properties in RCo12B6 compounds (R=Y, Gd, and Ho). J. Phys. Chem. B 101, 12 (2020).
50. Patel, A. K., Samatham, S. S., Suresh, K. G. Critical behavior, universality class and magnetotransport properties of Ni2MnIn. Mater. Res. Bull. 128, 8 (2020).
51. Li, N., Tian, Y., Zhao, J., Zhan, W., Du, J., Kong, L., Zhang, J., Zuo,W. Ultrafast selective capture of phosphorus from sewage by 3D Fe_3O_4@ZnO via weak magnetic field enhanced adsorption. Chem. Eng. J. 341, 289–297 (2018a).
52. Zielinski, M., Rusanowska, P., Debowski, M., Hajduk, A. Influence of static magnetic field on sludge properties. Sci. Total Environ. 625, 738–742 (2018).
53. Krzemieniewski, M., Debowski, M., Janczukowicz, W., Pesta, J. Effect of the constant magnetic field on the composition of dairy wastewater andwastewater and domestic sewage. Pol. J. Environ. Stud. 13, 45–53 (2004).
54. Gamzatov, A. G., Batdalov, A. B., Aliev, A. M., Yen, P. D. H., Gudina, S. V., Neverov, V. N., Thanh, T. D., Dung, N. T., Yu, S. C., Kim, D. H., Phan, M. H. Determination of the magnetocaloric effect from thermophysical parameters and their relationships near magnetic phase transition in doped manganites. J. Magn. Magn. Mater. 513, 167209 (2020).
55. Kiwi, J., Rtimi, S. Insight into the interaction of magnetic photocatalysts with the incoming light accelerating bacterial inactivation and environmental cleaning. Appl. Catal. B Environ. 281, 119420 (2020).
56. Piamba, J. F., Ortega, C., Hernández-Bravo, R., González, C. J. M., Tabares, J. A., Pérez Alcázar, G. A., Alvarado-Orozco, J. M. Theoretical and experimental study of FeSi on magnetic and phase properties. J. Appl. Phys. A (126): 849 (2020).
57. Hilou, E., Joshia, K., Biswal, L. S. Characterizing the spatiotemporal evolution of paramagnetic colloids in time-varying magnetic fields with Minkowski functionals. Soft Matter 16, 8799–8805 (2020).
58. Vick, W. S. Magnetic fluid conditioning. Proceedings of the 1991 Speciality Conference on Environmental Engineering. American Society of Civil Engineers, Reston, VA (1991).
59. Omar, A. H., Muda, K., Toemen, S., Sulaiman, S. F., Zaidi, N. S., Affam, A. C. Study on the effect of a static magnetic field in enhancing initial state of biogranulation. Aqua. jws. 67 (5), 484–489 (2018).
60. Virgen, M. D. R. M., Vázquez, O. F. G., Montoya, V. H., Gómez, R. T. Removal of heavy metals using adsorption processes subject to an external magnetic field (heavy metals) (2018).
61. Hattori, S., Watanabe, M., Osono, H., Togii, H., Sasaki, K. Effects of an external magnetic field on the flock size and sedimentation of activated sludge. World J. Microbiol. Biotechnol. 17, 833–838 (2001).
62. Wang, X. H., Diao, M. H., Yang, Y., Shi, Y. J., Gao, M. M., Wang, S. G. Enhanced aerobic nitrifying granulation by static magnetic field. Bioresour. Technol. 110, 105–110 (2012).
63. Yilin, W., Xin, G., Jianing, Q., Xing, G., Yang, L., Zhao, C., Wu, P., Zhao, F., Hua, B., Hu, Y. Application of magnetic fields to wastewater treatment and its mechanisms: A review. Science of the Total Environment 773, 145476 (2021).
64. Gqebe, S., Rodriguez-Pascual, M., Lewis, A. A modification of the zeta potential of copper sulphide by the application of a magnetic field in order to improve particle settling. J. South. Afr. Inst. Min. Metall. 116, 575–580 (2016).
65. Jelodari, I., Nikseresht, A. H. Effects of Lorentz force and induced electrical field on the thermal performance of a magnetic nanofluid-filled cubic cavity. J. Mol. Liq. 252, 296–310 (2018).

66. Udagawa, C., Ueno, M., Hisaki, T., Maeda, M., Maki, S., Morimoto, S., Tanimoto, Y. Magnetic field effects on electroless deposition of lead metal-Lorentz force effects. Bull. Chem. Soc. Jpn. 91, 165–172 (2018)

67. Huang, H. J., Wang, Y. H., Chau, Y. F. C., Chang, H. P., Wu, J. C. S. Magnetic field enhancing photocatalytic reaction in micro optofluidic chip reactor. Nanoscale Res. Lett. 14, 7 (2019).

68. Hao, X. L., Liu, H., Zhang, G. S., Zou, H., Zhang, Y. B., Zhou, M. M., Gu, Y. C. Magnetic field assisted adsorption of methyl blue onto organo-bentonite. Appl. Clay Sci. 55, 177–180 (2012).

69. Naletova, V. A., Turkov, V. A., Tyatyushkin, A. N., 2005. Spherical body in a magnetic fluid in uniform electric and magnetic fields. J. Magn. Mater. 289, 370–372 (2005).

70. Hao, X. L., Liu, H., Zhang, G. S., Zou, H., Zhang, Y. B., Zhou, M. M., Gu, Y. C. Magnetic field assisted adsorption of methyl blue onto organo-bentonite. Appl. Clay Sci. 55, 177–180 (2012).

70. Ma, J., Ma, Y., Yu, F., Dai, X. H. Rotating magnetic field-assisted adsorption mechanism of pollutants on mechanically strong sodium alginate/graphene/Lcysteine beads in batch and fixed-bed column systems. Environ. Sci. Technol. 52, 13925–13934 (2018a).

71. Li, J. X., Qin, H. J., Zhang, W. X., Shi, Z., Zhao, D. Y., Guan, X. H. Enhanced Cr (VI) removal by zero-valent iron coupled with weak magnetic field: role of magnetic gradient force. Sep. Purif. Technol. 176, 40–47 (2017).

72. Pirsaheb, M., Moradi, S., Shahlaei, M., Wang, X., Farhadian, N. Simultaneously implement of both weak magnetic field and aeration for ciprofloxacin removal by Fenton-like reaction. J. Environ. Manag. 246, 776–784 (2019).

73. Velev, O. D., Bhatt, K. H., 2006. On chip manipulation and assembly of colloidal particles by electric fields. J. R. Soc. Med. 100, 38–750 (2006).

74. Rivera, F. L., Palomares, F. J., Herrasti, P., Mazario, E. Improvement in heavy metal removal from wastewater using an external magnetic inductor. Nanomaterials. 9, 15 (2019).

75. Zhang, G. K., Liu, Y., Xie, Y., Yang, X., Hu, B., Ouyang, S. X., Liu, H. X., Wang, H. Y. Zinc adsorption on Narectorite and effect of static magnetic field on the adsorption. Applied Clay Ence. 29, 15–21 (2005).

76. Qin, Y., Dong, B., Li, W. Z., 2020. Experimental study of the frosting characteristic of water on a cold surface in the magnetic field. Exp. Thermal Fluid Sci. 114, 12 (2020).

77. Chang, K. T., Cheng, I. W. An investigation into the structure of aqueous NaCl electrolyte solutions under magnetic fields. Comput. Mater. Sci. 43, 1048–1055 (2008).

78. Lee, S. H., Takeda, M., Nishigaki, K. Gas–liquid interface deformation of flowing water in gradient magnetic field–influence of flow velocity and NaCl concentration. Jpn. J. Appl. Phys. 42, 1828–1833 (2014).

79. Chang, K. T., Weng, C. I. The effect of an external magnetic field on the structure of liquid water using molecular dynamics simulation. J. Appl. Phys. 100, 2923 (2006).

80. Gokon, N., Shimada, A., Kaneko, H., Tamaura, Y., Ito, K., Ohara, T. Magnetic coagulation and reaction rate for the aqueous ferrite formation reaction. J. Magn. Mater. 238, 47–55 (2002).

81. Hakobyan, S. N., Ayrapetyan, S. N. A study of specific electrical conductivity of water by the action of constant magnetic field, electromagnetic field, and low frequency, mechanical vibrations. Biophysics. 50, 265–270 (2005).

82. Holysz, L., Szczes, A., Chibowski, E. Effects of a static magnetic field on water and electrolyte solutions. J. Colloid Interface Sci. 316, 996–1002 (2007).

83. Iwasaka, M., Ueno, S. Structure of water molecules under 14 T magnetic field. J. Appl. Phys. 83, 6459–6461 (1998).
84. Ishii, K., Yamamoto, S., Yamamoto, M., Nakayama, H. Relative change of viscosity of water under a transverse magnetic field of 10 T is smaller than 104. Chem. Lett. 34, 874–875 (2005).
85. Levitt, M., Hirshberg, M., Sharon, R., Laidig, K. E., Daggett, V. Calibration and testing of a water model for simulation of the molecular dynamics of proteins and nucleic acids in solution. Phys. Chem. B 101, 5051–5061 (1997).
86. Inaba, H., Saitou, T., Tozaki, K., Hayashi, H. Effect of the magnetic field on the melting transition of H2O and D2O measured by a high resolution and supersensitive differential scanning calorimeter. J. Appl. Phys. 96, 6127–6132 (2004).
87. Cai, R., Yang, H., He, J., Zhu, W. The effects of magnetic fields on water molecular hydrogen bonds. J. Mol. Struct. 938, 15–19 (2009).
88. Toledo, E. J. L., Ramalho, T. C., Magriotis, Z. M. Influence of magnetic field on physical chemical properties of liquid water. Insights from experimental and theoretical models. J. Mol. Struct. 888, 409–415 (2008).
89. Wang, Y., Zhang, B., Gong, Z., Gao, K., Zhang, J. The effect of a static magnetic field on the hydrogen bonding in water using frictional experiments. J. Mol. Struct. 1052, 102–104 (2013).
90. Ma, J., Ma, Y., Yu, F. A novel one-pot route for large-scale synthesis of novel magnetic CNTs/Fe@C hybrids and their applications for binary dye removal. ACS Sustain. Chem. Eng. 6, 8178–8191 (2018b).
91. Niu, C., Geng, J. J., Ren, H. Q., Ding, L. L., Xu, K., Liang, W. H. The strengthening effect of a static magnetic field on activated sludge activity at low temperature. Bioresour. Technol. 150, 156–162 (2013).
92. Łebkowska, M., Rutkowska-Narozniak, A., Pajor, E., Tabernacka, A., Załęska-Radziwiłł, M. Impact of a static magnetic field on biodegradation of wastewater compounds and bacteria recombination. Environ. Sci. Pollut. Res. 25, 22571–22583 (2018).
93. Zielinski, M., Rusanowska, P., Debowski, M., Hajduk, A. Influence of static magnetic field on sludge properties. Sci. Total Environ. 625, 738–742 (2018).
94. Liu, S., Yang, F., Meng, F., Chen, H., Zheng, G. Enhanced anammox consortium activity for nitrogen removal: impacts of static magnetic field. J. Biotechnol. 138, 96–102 (2008).
95. Tan, L., Shao, Y. F., Mu, G. D., Ning, S. X., Shi, S. N. Enhanced azo dye biodegradation performance and halotolerance of Candida tropicalis SYF-1 by static magnetic field (SMF). Bioresour. Technol. 295, 122283 (2020b).
96. Dini, L., Abbro, L. Bioeffects of moderate-intensity static magnetic fields on cell cultures. Micron. 36, 195–217 (2005).
97. Pospisilova, D., Schreiberova, O., Jirku, V., Lederer, T. Effects of magnetic field on phenol biodegradation and cell physiochemical properties of rhodococcus erythropolis. Bioremediat. J. 19, 201–206 (2015).
98. Hu, Q., Joshi, R. P., Miklavcic, D. Calculations of cell transmembrane voltage induced by time-varying magnetic fields. IEEE Trans. Plasma Sci. 48, 1088–1095 (2020b).
99. Lan, H. X., Chen, R., Ma, P., Zhang, H., Lan, S. H., Wang, Y. D. Cultivation and characteristics of microaerobic activated sludge with weak magnetic field. Desalin. Water Treat. 53, 27–35 (2015).
100. Qian, J. Y., Zhang, M., Dai, C. H., Huo, S. H., Ma, H. L., 2020. Transcriptomic analysis of Listeria monocytogenes under pulsed magnetic field treatment. Food Res. Int. 133, 10 (2020).

101. Koch, C. L. M. B., Sommarin, M., Persson, B. R. R., Salford, L.G., Eberhardt, J. L. Interaction between weak low frequency magnetic fields and cell membranes. Bioelectromagnetics. 24, 395–402 (2003).

102. Hughes, S., McBain, S., Dobson, J., El Haj, A. J. Selective activation of mechanosensitive ion channels using magnetic particles. J. R. Soc. Interface 5, 855–863 (2008).

103. Panagopoulos, D. J., Karabarbounis, A., Margaritisa, L. H. Mechanism for action of electromagnetic fields on cells. Biochem. Biophys. Res. Commun. 298, 95–102 (2002).

104. Jovanova-Nesic, K., Eric-Jovicic, M., Spector, N. H. Magnetic stimulation of the brain increase Na+, K+-ATPase activity decreased by injection of AlCl3 into nucleus basalis magnocellularis of rats. Int. J. Hyperth. 116, 681–695 (2006).

105. Rosen, A. D. Effect of a 125mT static magnetic field on the kinetics of voltage activated Na+ channels in GH3 cells. Bioelectromagnetics. 24, 517–523 (2003).

106. Lin, S. L., Su, Y. T., Feng, S. W., Chang, W. J., Fan, K. H., Huang, H. M. Enhancement of natural killer cell cytotoxicity by using static magnetic field to increase their viability. Electromagn. Biol. Med. 38, 131–142 (2019).

107. Wang, Z., Sarje, A., Che, P. L., Yarema, K. J. Moderate strength (0.23–0.28 T) static magnetic fields (SMF) modulate signaling and differentiation in human embryonic cells. Bmc Genomics 10, 356 (2009).

108. Pazur, A., Schimek, C., Galland, P. Magnetoreception in microorganisms and fungi. Cent. Eur. J. Biol. 2, 597–659 (2007).

109. Letuta, U. G., Avdeeva, E. I., Berdinsky, V. L. Magnetic field effects in bacteria E. coli in the presence of Mg isotopes. Russ. Chem B. 63, 1102–1106 (2014).

110. Kamel, F. H., Saeed, C. H., Qader, S. S. The effects of magnetic fields on some biological activities of Pseudomonas aeruginosa. Diyala J Med. 5, 29–35 (2013).

111. Zheng, M. Q., Su, Z. G., Ji, X.Y., Ma, G. H., Wang, P., Zhang, S. P. Magnetic field intensified bienzyme system with in situ cofactor regeneration supported by magnetic nanoparticles. J. Biotechnol. 168, 212–217 (2013).

112. Wasak, A., Drozd, R., Jankowiak, D., Rakoczy, R. Rotating magnetic field as tool for enhancing enzymes properties-laccase case study. Sci. Rep. 9, 9 (2019b).

113. Jung, J., Sofer, S. Enhancement of phenol biodegradation by south magnetic field exposure. J. Chem. Technol. Biotechnol 70, 299–303 (1997).

114. Magazu, S., Calabro, E. Studying the electromagnetic-induced changes of the secondary structure of bovine serum albumin and the bioprotective effectiveness of trehalose by fourier transform infrared spectroscopy. J. Phys. Chem. B 115, 6818–6826 (2011).

115. Fraga, F. C., Valerio, A., Oliveira, V. A., Luccio, M., Oliveira, D. Effect of magnetic field on the Eversa (R) Transform 2.0 enzyme: enzymatic activity and structural conformation. Int. J. Biol. Macromol. 122, 653–658 (2019).

116. Buchachenko, A., Kuznetsov, D., 2014. Magnetic control of enzymatic phosphorylation. J. Phys. Chem. B 4, 2 (2014).

117. Logeshwaran, P., Krishnan, K., Naidu, R., Megharaj, M. Purification and characterization of a novel fenamiphos hydrolysing enzyme from Microbacterium esteraromaticum MM1. Chemosphere. 252, 126549 (2020).

118. Zhang, X.,Wan, L., Li, L., Xu, Z., Su, J., Li, B., Huang, J. Effects ofmagnetic fields on the enzymatic synthesis of naringin palmitate. RSC Adv. 8, 13364–13369 (2018).

119. Zhang, X., Yarema, K., Xu, A., 2017. Biological effects of static magnetic fields ‖ impact of static magnetic fields (SMFs) on cells. Springer Singapore. 34, 758–763 (2017).

120. Mu, B., Wang, A. Q., 2015. One-pot fabrication of multifunctional superparamagnetic attapulgite/Fe_3O_4/polyaniline nanocomposites served as an adsorbent and catalyst support. J. Mater. Chem. 3, 281–289 (2015).

121. Xiong, T., Yuan, X. Z., Cao, X. Y., Wang, H., Jiang, L. B.,Wu, Z. B., Liu, Y. Mechanistic insights into heavy metals affinity in magnetic MnO_2@Fe_3O_4/ poly(mphenylenediamine) coreshell adsorbent. Ecotox. Environ. Safe. 192, 9 (2020).

122. Li, W. S., Zheng, A. M., Liu, J., Li, G. Y., Wang, W. Y., Yang, Y. Q. Gas-assisted low-field magnetic separation for efficient recovery of contaminants-loaded magnetic nanoparticles from large volume water solution. Sep. Purif. Technol. 248, 117016 (2020b).

123. Szatylowicz, E., Skoczko, I. Magnetic field usage for the removal of iron by filtration assisted different filter materials. Proceedings. 16, 6 (2019).

124. Rajczykowski, K., Loska, K. Stimulation of heavy metal adsorption process by using a strong magnetic field. Water Air Soil Pollut. 229, 7 (2018).

125. Tireli, A. A., Marcos, F. C. F., Oliveira, L. F., Guimaraes, I. D., Guerreiro, M. C., Silva, J. P. Influence of magnetic field on the adsorption of organic compound by clays modified with iron. Appl. Clay Sci. 97–98, 1–7 (2014).

126. Wang, M. J., Zhang, J., Zhao, H. D., Deng, W. N., Lu, J. F., Ye, Q. Enhancement of oxidation capacity of ZVI/Cu^{2+}/PMS systems by weak magnetic fields. Desalin. Water Treat. 161, 260–268 (2019).

127. Aigbe, U. O., Khenfouch, M., Ho, W. H., Maity, A., Vallabhapurapu, V.S., Hemmaragala, N.M. Congo red dye removal under the influence of rotatingmagnetic field by polypyrrole magnetic nanocomposite. Desalin. Water Treat. 131, 328–342 (2018a).

128. An, L., Meng, Y., Wang, T., Xiong, C., Yan, Z. X., Xu, Z. H. Highly efficient and easily recoverable Ag_3PO_4-TiO_2-Fe_2O_3 magnetic photocatalyst with wide spectral range for water treatment. Russ. J. Phys. Chem. A 94, 1067–1072 (2020).

129. Zheng, X. Y., Zheng, H. L., Xiong, Z. K., Zhao, R., Liu, Y. Z., Zhao, C., Zheng, C. F. Novel anionic polyacrylamide-modify-chitosan magnetic composite nanoparticles with excellent adsorption capacity for cationic dyes and pH-independent adsorption capability for metal ions. Chem. Eng. J. 392, 15 (2020).

130. Tan, L., He, M. Y., Song, L., Fu, X. M., Shi, S. N. Aerobic decolorization, degradation and detoxification of azo dyes by a newly isolated salt-tolerant yeast Scheffersomyces spartinae TLHS-SF1. Bioresour. Technol. 203, 287–294 (2016).

131. Song, L., Shao, Y. F., Ning, S. X., Tan, L. Performance of a newly isolated salt-tolerant yeast strain Pichia occidentalis G1 for degrading and detoxifying azo dyes. Bioresour. Technol. 233, 21–29 (2017).

132. Wang, X. H., Wang, Y. M., Ning, S. X., Shi, S. N., Tan, L. Improving azo dye decolorization performance and halotolerance of Pichia occidentalis A2 by static magnetic field and possible mechanisms through comparative transcriptome analysis. Front. Microbiol. 11, 712 (2020).

133. Tan, L., Mu, G. D., Shao, Y. F., Ning, S. X., Shi, S. N. Combined enhancement effects of static magnetic field (SMF) and a yeast Candida tropicalis SYF-1 on continuous treatment of Acid Red B by activated sludge under hypersaline conditions. J. Chem. Technol. Biotechnol. 95, 840–849 (2020a).

134. Wasak, A., Drozd, R., Jankowiak, D., Rakoczy, R. The influence of rotating magnetic field on bio-catalytic dye degradation using the horseradish peroxidase. Biochem. Eng. J. 147, 81–88 (2019a).

135. Gu, S. G., Lian, F., Yan, K. J., Zhang, W. Application of polymeric ferric sulfate combined with cross frequency magnetic field in the printing and dyeing wastewater treatment. Water Sci. Technol. 80, 1562–1570 (2019).

136. Ma, J., Ma, Y., Yu, F., Dai, X. H. Rotating magnetic field-assisted adsorption mechanim of pollutants on mechanically strong sodium alginate/graphene/Lcysteine beads in batch and fixed-bed column systems. Environ. Sci. Technol. 52, 13925–13934 (2018a).

137. Lipus, L. C., Hamler, A., Buchmeister, B., Gorsek, A. Magnetic treatment for amelioration of wastewater biodegradation. DAAAM International Scientific Book. 9, 097–106 (2018).

138. Kamariah, M. S. Subsurface Flow and Free Water Surface Flow Constructed Wetland with Magnetic Field for Leachate Treatment. University of Teknologi, Malaysia, Dissertation (2006).

139. Wahid, Z. A., Othman, F., Sohaili, J. Electromagnetic technology on sewage treatment. Malaysian Journal of Civil Engineering. 13, 11–21 (2001).

140. Bao, T., Damtie, M. M., Yu, Z. M., Liu, Y., Jin, J., Wu, K., Deng, C.X., Wei, W., Wei, X. L., Ni, B. Green synthesis of Fe_3O_4@Carbon filter media for simultaneous phosphate recovery and nitrogen removal from domestic wastewater in biological aerated filters. ACS Sustain. Chem. Eng. 7, 16698–16709 (2019).

141. Xu, D., Ji, H. M., Ren, H. Q., Geng, J. J., Li, K., Xu, K. Inhibition effect of magnetic field on nitrous oxide emission from sequencing batch reactor treating domestic wastewater at low temperature. J. Environ. Sci. 87, 205–212 (2020).

142. Li, B., Tan, G., Wang, M., Zhang, D., Liu, W. Electric fields and local magnetic field enhance Ag-$BiVO_4$-MnO_x photoelectrochemical and photocatalytic performance. Appl. Surf. Sci. 511, 10 (2020a).

143. Lu, Y. K., Ren, B. Y., Chang, S. C., Mi, W. B., He, J., Wang, W. Achieving effective control of the photocatalytic performance for $CoFe_2O_4$/MoS_2 heterojunction via exerting external magnetic fields. Mater. Res. Lett. 260, 4 (2020).

144. Shi, L., Wang, X. Z., Hu, Y. W., He, Y. R. Investigation of photocatalytic activity through photo-thermal heating enabled by Fe_3O_4/TiO_2 composite under magnetic field. Sol. Energy 196, 505–512 (2020).

145. Zhao, J. H., Li, N., Yu, R. X., Zhao, Z. W., Nan, J. Magnetic field enhanced denitrification in nitrate and ammonia contaminated water under 3D/2D Mn_2O_3/g-C_3N_4 photocatalysis. Chem. Eng. J. 349, 530–538 (2018).

146. Vinosel, V. M., Anand, S., Janifer, M. A., Pauline, S., Dhanavel, S., Praveena, P., Stephen, A., 2019. Enhanced photocatalytic activity of Fe_3O_4/SnO_2 magnetic nanocomposite for the degradation of organic dye. J. Mater. Sci.-Mater. Electron. 30, 9663–9677

147. Kakavandi, B., Takdastan, A., Jaafarzadeh, N., Azizi, M., Mirzaei, A., Azari, A., 2016. Application of Fe_3O_4@C catalyzing heterogeneous UV-Fenton system for tetracycline removal with a focus on optimization by a response surface method. J. Photochem. Photobiol. A-Chem. 314, 178–188.

148. Li, Q. L., Kong, H., Li, P., Shao, J. H., He, Y. L., 2019. Photo-Fenton degradation of amoxicillinvia magnetic TiO_2-graphene oxide-Fe_3O_4 composite with a submerged magnetic separation membrane photocatalytic reactor. J. Hazard. Mater. 373, 437–446.

149. Barbosa, I. A., Zanatta, L. D., Espimpolo, D. M., Silva, D. L., Nascimento, L. F., Zanardi, F. B., De, S. F. P. C., Serra, O. A., Iamamoto, Y. Magnetic diatomite (Kieselguhr)/Fe_2O_3/TiO_2 composite as an efficient photo-Fenton system for dye degradation. Solid State Sci. 72, 14–20 (2017).

150. Cai, H., Zhao, T. C., Ma, Z. C., Liu, J. Z. Efficient removal of metronidazole by the Photo Fenton process with a magnetic Fe_3O_4@PBC composite. J. Environ. Eng.-ASCE. 146(7): 04020056 (2020).

151. Cai, C., Zhang, Z., Jin, L., Ni, S., Dionysiou, D. D. Visible light-assisted heterogeneous Fenton with $ZnFe_2O_4$ for the degradation of Orange II in water. Appl. Catal. BEnviron. 182, 456–468 (2016).

152. Hu, L., He, H., Xia, D., Huang, Y., Xu, J., Li, H., He, C., Yang, W., Shu, D., Wong, P. K. ACS Appl. Mater. Inter. 10, 18693–18708 (2018).

153. Xu, L., Wang, J. Magnetic nano scaled Fe_3O_4/CeO_2 composite as an efficient Fenton like heterogeneous catalyst for degradation of 4-Chlorophenol. Environ. Technol. 46, 10145–10153 (2012).

154. Li, N., Tian, Y., Zhao, J., Zhan, W., Du, J., Kong, L., Zhang, J., Zuo, W. Ultrafast selective capture of phosphorus from sewage by 3D Fe_3O_4@ZnO via weak magnetic field enhanced adsorption. Chem. Eng. J. 341, 289–297 (2018a).

155. Morrow, A. C., Dunstan, R. H., King, B. V., Roberts, T. K. Metabolic effects of static magnetic fields on Streptococcus Pyogenes. Bioelectromagnetics. 28, 439–445 (2007).

156. Bajpai, I., Saha, N., Basu, B. Moderate intensity static magnetic field has bactericidal effect on E. coli and Sepidermidis on sintered hydroxyapatite. J. Biomed. Mater. Res. Part B. 100B, 1206–1217 (2012).

157. Ferrada, P., Rodriguez, S., Serrano, G., Miranda, O. C., Maureira, A., Zapata, M. An analytical experimental approach to quantifying the effects of static magnetic fields for cell culture applications. Appl. Sci.-Basel. 10, 22 (2020).

158. Aigbe, U. O., Das, R., Ho, W. H., Srinivasu, V., Maity, A. A novel method for removal of Cr(VI) using poly pyrrole magnetic nanocomposite in the presence of unsteady magnetic fields. Sep. Purif. Technol. 194, 377–387 (2018b).

159. Messiha, H. L., Wongnate, T., Chaiyen, P., Jones, A. R., Scrutton, N. S. Magnetic field effects as a result of the radical pair mechanism are unlikely in redox enzymes. J. R. Soc. Interface 12, 10 (2015).

160. Lopez, S. L. F., Moreno-Virgen, M. R., Hernández-Montoya, V., Montes-Morán, M. A., Tovar, G. R., Rangel, V. N. A., Pérez, C. M. A., Esparza, G. M. S. Effect of an external magnetic field applied in batch adsorption systems: removal of dyes and heavy metals in binary solutions. J. Mol. Liq. 269, 450–460 (2018).

7 Photo-Catalytic Reactor
A Modeling Approach

Brajesh Kumar[1], Vijay Singh[2], Pradeep Kumar[2], and Krunal M. Gangawane[3]*
[1]Department of Chemical Engineering, National Institute of Technology Srinagar, J&K, India
[2]Department of Chemical Engineering, Institute of Engineering & Technology Lucknow, U.P., India
[3]Department of Chemical Engineering, National Institute of Technology Rourkela, Odisha, India
*Corresponding Author: brajesh10iitr@gmail.com

CONTENTS

7.1 INTRODUCTION

Currently, valuable fuels procured from conventional sources of energy are depleting due to exorbitant consumption for the rapid industrial growth in which the requirement of energy is the primary need to generate electricity and the production of various basic and advanced things for human beings [1]. Moreover, continuous utilization of limited natural resources is raising the concentration of greenhouse gas CO_2 in the atmosphere. Because of these reasons, the serious repercussions related with the environment and energy is waiting in the near future. Therefore, the search for an alternative source is essential to fulfil this energy need and replacement of conventional sources of energy for the sustainable progress of society. Solar-to-fuel conversion technologies are getting more research attention as utilization of renewable sources of energy, which will definitely reduce the dependency on conventional fuels [2]. Among them, photocatalysis has evolved as a promising technology for energy and environmental safety due to its potential for solar energy utilization and storage. It can provide benefits in the fields of energy, environment, pollution control, and value-added chemical synthesis.

Photocatalysis is a photo-induced process for oxidation that requires participation of a photocatalyst (mostly TiO_2), ultra-visible incident solar radiation, and oxygen to decompose organic materials released in liquid wastes (such as dyes, surfactants, and organic pesticides) or gaseous wastes in the form of carbon dioxide emissions along with inorganic acid and water [3]. Oxidizing agents can be replaced with ozone or hydrogen peroxide as per availability. However, photocatalysis is a very popular and effective process for low pollutant concentration. In the photocatalysis process, a photocatalyst generates electrons or holes on its surface due to absorption of light as ultra-visible incident solar radiation [4].

The general mechanism of photocatalytic reaction can be represented as follows [2]:

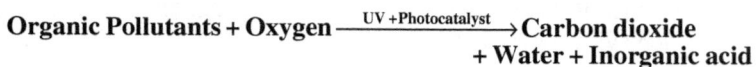

$$\textbf{Organic Pollutants} + \textbf{Oxygen} \xrightarrow{\text{UV +Photocatalyst}} \textbf{Carbon dioxide} + \textbf{Water} + \textbf{Inorganic acid}$$

The great challenge in photocatalysis is the development and implementation of this technology at industrial scale to get more advantages.

Modeling of a chemical reactor is based on fundamental conservation balance equations of momentum, mass, and thermal energy, whereas a photocatalytic reactor requires radiant energy distribution within the reaction space. Photocatalytic reactors are mainly used to carry out photocatalytic reactions, preferably heterogeneous reactions. In modeling a photocatalytic reactor, the evaluation of radiant photon absorption rate inside the photocatalytic reactor is necessary along with the conservation of a mass balance equation, which will give the idea of intermediate species along with desired components in the system. Due to fixed temperature (mostly room temperature), there is no need to take account of the thermal energy balance equation. The reaction rate of each feasible component participated in the reaction system is necessarily required to solve the mass balance equations. Therefore, kinetic rate expressions obtained from a stoichiometric approach are required for the photocatalytic reaction system. The first step of a kinetic scheme is the activation step in which absorption of heat radiation is done by the photocatalyst for the formation of electron-hole pairs [3,4]. The absorption rate of photons consists of the volumetric rate for the reactions with the catalyst in the aqueous solution and the surface rate for the reaction system with catalyst particles deposited on their inert support. Therefore, a clear picture is necessary for kinetic modeling of photocatalytic reactors [5].

This kind of complex mathematical modeling of photocatalytic reactors depends on sophisticated advanced software tools. The available CFD software tools for carrying out study for such complex modeling of photocatalytic reaction systems are ANSYS Fluent and COMSOL Multiphysics. Other open-source CFD softwares are OpenFOAM and FEATFLOW [6,7]. The present chapter considers a concise approach for modeling the photocatalytic reactor in which characteristics, mathematical formulation, and the effects of various parameters are discussed in the forthcoming sections.

7.2 REACTOR CHARACTERISTICS

A chemical reactor modeling is based on the knowledge of fundamental conservation laws of momentum, mass (including multicomponents), and energy (thermal). It is also dependent on several characteristics like scale of production (batch, semi-batch, and continuous flow), mode of operation (isothermal, non-isothermal, adiabatic), and contacting pattern between phases (e.g., packed bed, fluidized bed, and bubble column). On the other hand, modeling of a photocatalytic reactor also requires the distribution of radiant energy within the reactor volume along with aforementioned characteristics of a chemical reactor according to the variety of reaction [7]. In a photocatalytic reactor, the study of radiation field is done and the evaluation of the photon absorption rate is also analyzed. The characterization of a photocatalytic reactor depends upon some essential factors such as structure/shape (cylindrical and rectangular), fluid dynamics, and other factors. The discussion regarding factors that characterized the photocatalytic reactor is as follows:

7.2.1 SHAPE/STRUCTURE OF PHOTOCATALYTIC REACTOR

Structure or shape plays an important role in the functioning of a photocatalytic reactor and depends upon flow rate (volumetric/mass/molar), reactor volume, characteristics of contaminants as dosage, concentration, etc., which directly affects the efficiency of the photocatalytic reactor. The structure that is used to fabricate the photocatalytic reactor is categorized into two shapes as cylindrical and rectangular.

7.2.1.1 Cylindrical Photocatalytic Reactor

In a cylindrical photocatalytic reactor, lamps used for obtaining thermal radiation for occurring photoreaction are organized in the form of a circle in an attempt to obtain a proper distribution of thermal energy inside the reactor. The catalysts used in the cylindrical photocatalytic reactor are dispersed into volume of liquid reagent.

The reactor system setup consists of a storage vessel attached with an aeration pump and mixer, recirculation pump, sampling and control valves, flow meter, and a control system. The process of aeration is done in the storage vessel so that saturation of oxygen can be settled to the formation of oxidative radicals. This type of reactor possesses the benefits of radial flow distribution to raise the diffusion of homogeneous activity of the photocatalyst.

Cylindrical photocatalytic reactors are used to manage high reactor volume and high flow rates as well. Apart from numerous advantages of this kind of shape for the reactor has limitation that it is not capable enough to predict the behavior of reaction in an adequate manner and changes in the contaminant types, catalyst activity, and related chemistry.

7.2.1.2 Rectangular Photocatalytic Reactor

The rectangular shape helps to get radiation from the light source inside the reactor in any manner such as central, horizontal, and vertical directions/positions at any corner. This reactor system setup also contains several components like aeration and recirculation pumps, storage vessel, mixer, sampling and control valves, flowmeter, and a control system. Everything would be the same as the cylindrical photoreactor except the shape and different distribution of radiant energy inside the reactor. The immobilized type of the photocatalyst is utilized to manage flow rate and homogeneous diffusion through the reactor volume to this type of geometry as rectangular shape.

This type of photocatalytic reactor is chosen to generate the maximum number of oxidative radicals to remove organic impurities from wastewater. Due to simplicity in shape/geometry, this type of reactor can easily be fabricated for small-scale applications at an indoor level and large-scale operation at an outdoor level. This kind of reactor also has the issue with the optimization of flow rate and radiation distribution in a uniform manner.

7.2.2   Fluid Dynamics

The design of a photocatalytic reactor is a tedious task because of the involvement of hydrodynamics of the reactor. Fluid dynamics of any type of reactor focuses on several challenges and their possible solution, such as modification of the photocatlyst to utilize more solar spectrum, to handle solid-liquid separation, to study fluid flow and the development of a proper design to enhance the efficiency of the photoreactor. The detailed study of fluid dynamics of the photoreactor is discussed as modeling the photoreactor in subsequent sections in the present chapter.

7.2.3   Other Factors

There are many other factors except as previously discussed, on which basis the characterization of the photocatalytic reactor depends, such as the type of reaction, type of catalysts, scale of operation, mode of operation, flow of reagents and products, contacting patters between phases.

7.3   MATHEMATICAL FORMULATION

The photocatalytic reactor can be divided into two kinds of systems, i.e., immobilized photocatalytic reaction systems with the presence of catalysts on the surface and slurry-type photocatalytic reaction systems where catalysts act as suspended particles. Immobilized photocatalytic reaction systems are known as versatile reactors because of multitudinous names according to the suitability as optical fiber, flat plate, packed bed, monolith, multi-annular, fixed bed, annular venture, Taylor vortex, etc. These reactors do not prefer post treatment for costly catalytic recovery from nano- to micro-sized materials. Slurry reactors have also many types as thin film slurry (TFS) reactors, moving bed reactors, trickle bed reactors, fluidized beds, and externally illuminated aerated tank (EIART) reactors. The performance of a slurry-type photocatalytic reactor is better than an immobilized photocatalytic reactor because of the better light exposure and high mass transfer coefficient to the catalyst particle [5–7].

For different applications, modeling of ultraviolate photocatalytic reactors has been developed to see their performance in various aspects. Modeling of the photocatalytic reactor via computational fluid dynamics consists of four steps: hydrodynamic modeling, lamp source modeling, radiation transport modeling, and kinetic modeling. The equations, data, and correlation in these steps are somehow connected to each other as per system availability whether the system is immobilized or slurry reactor [7].

Hydrodynamic modeling comprises multiphase flow in which species transport and catalyst phase distribution are significant parts. The modeling equations in the hydrodynamic modeling step is followed by lamp source modeling, which has many kinds of source models as the line source model, surface source model, and volume source model. In the next step, radiation transport modeling come into existence and provides correlation data, system-wise equations as immobilized systems and slurry systems. Immobilized systems are based on surface irradiance connected with the earlier step of species transport in hydrodynamic modeling and are also related with the forthcoming step of the surface reaction in kinetic modeling. Slurry systems are based on incident radiation in connection with catalyst phase distribution and the local volumetric rate of energy absorption. In this step, further optical properties and the local volumetric rate of energy absorption are studied in which optical properties follow the path of phase distribution and on the other side, the local volumetric rate of energy absorption are connected with species transport and volumetric reaction. In the last step, kinetic modeling is studied for the immobilized system (surface reaction) and the slurry system (volumetric reaction) as per the choice of the reaction system.

7.3.1   Step I: Hydrodynamic Modeling

For single phase flow, hydrodynamics of the photocatalytic reactor is studied via the Naiver–Stokes equation. In the case of multiphase flow, two models are commonly used as Eulerian–Eulerian (E–E) and Eulerian–Lagrangian (E–L) models [7]. These two models are discussed in detail below.

7.3.1.1   Eulerian–Eulerian (E–E) Model

This model is based on conservation of momentum and mass equations for multi-phase flow. The formulation of volume fraction for n phases (multiphases=total no. of phases) present in the system is written as

$$\sum_{k=1}^{n} \alpha_k = 1 \tag{7.1}$$

where
  k = phases, k = 1, 2,..., n
  α_k = volume fraction of phase k.

The continuity equation for multiphase flow can be presented as

$$\frac{\partial(\alpha_k \rho_k)}{\partial t} + \nabla \alpha_k \rho_k U_k = \sum_{p=1, p\neq k}^{n} S_{pk} \tag{7.2}$$

where
  S_{pk} = mass transfer rate from phase p to phase k
  ρ_k = density of phase k
  U_k = velocity of fluid in phase k.

For phase k, the momentum balance equation can be introduced as

$$\frac{\partial(\alpha_k \rho_k U_k)}{\partial t} + \nabla \alpha_k \rho_k U_k U_k = -\alpha_k \nabla_p - \nabla\left(\alpha_k \tau_k\right) + \alpha_k \rho_k g + F_k + F_g \tag{7.3}$$

where
  F_k = interphase momentum exchange terms between phase k and all other phases in the system.
  F_g = interphase momentum exchange terms between solid phases (granular multi-phase flows).

Now, interphase coupling in Eq. (7.3) can be understood by following relation in terms of F_k :

$$F_k = \sum_{q=1}^{n} K_{kq}(U_q - U_k) \tag{7.4}$$

where
  K_{kq} = interphase momentum exchange coefficient
  U_q = velocity of fluid in phase q.

7.3.1.2   Eulerian-Lagrangian (E–L) Model

This model is based on the trajectories of the dispersed phase particles and related equation of motion in a continuous way, which can be explained by the following equation:

$$m_p \frac{dU_p}{dt} = F_p + F_G + F_D + F_{VM} + F_L + F_H \qquad (7.5)$$

where
  m_p = mass vector of particle
  U_p = velocity vector of particle.

F_p, F_G, F_D, F_{VM}, F_L, and F_H are interphase momentum exchange forces due to pressure gradient, gravity, drag, virtual mass, lift, and Basset history, respectively.

7.3.2   STEP II: LAMP SOURCE MODELING

Illumination of thephotocatalytic reactor can be done by three possible routes as direct reflection from distance to the immobilized system, direct immersion of the lamp in the reaction mixture volume for slurry systems and both direct and indirect reflection of instruments into the reactors. The three basic models are used for lamp emission modeling, such as line source, surface source, and volume source [5,7].

The line source model provides a more accurate result while surface and volume source models include all dimensions of the heat radiation. The line source model is simple and takes less computational time. The linear source model works well according to the low ratio of lamp radius to inner wall radius and high ratio of inner wall radius to the length of lamp. Further, if the ratio of lamp radius to lamp length becomes low, the linear source approaches to volume source, especially in the case of the annular reactor. On the other side, the reactors with reflecting devices should follow the volume source model [7–9].

In the slurry photocatalytic reactor, an ultraviolate lamp is used to describe the surface source model. Ultraviolate absorption plays an important role in the modeling of slurry reactors. This absorption is negligible in quartz sleeve reactors as compared to the annular reactor. Therefore, the effect of calculation about the central lamp associated in the system cannot be added in the mathematical formulation of modeling of the photocatalytic reactor through the lamp emission model. Further, uniform distribution of heat energy over the whole length of the lamp cannot be considered for accurate results because of characterization of the surface of the lamp. In some cases, all three models produce somewhat similar results but with minimal difference. The supporting modeling equations for lamp emission source are given in Table 7.1 as below:

TABLE 7.1
Equations for Lamp Emission Source Model [7]

Model	Emission Type	Mathematical Expression	Eq.
Line Source	Specular emission	$E_\lambda = \dfrac{k_{\lambda 1}}{4\pi} \displaystyle\int_{-L}^{L} \dfrac{dh}{(r^2 + (z-h)^2)}$	(7.6)
	Diffuse emission	$E_\lambda = \dfrac{k_{\lambda 1}}{4\pi} \displaystyle\int_{-L}^{L} \dfrac{rdh}{(r^2 + (z-h)^2)^{3/2}}$	(7.7)
Surface Source	Specular emission	$E_\lambda = \dfrac{k_{\lambda 2}}{4\pi} \displaystyle\int_{h=-L}^{h=+L} \displaystyle\int_{\phi=-\pi/2}^{\phi=+\pi/2} \dfrac{Rd\phi dh}{[(rcos\Theta - Rcos\phi)^2 + (rsin\Theta - Rsin\phi)^2 + (z-h)^2]}$	(7.8)
	Diffuse emission	$E_\lambda = \dfrac{k_{\lambda 2}}{4\pi} \displaystyle\int_{h=-L}^{h=+L} \displaystyle\int_{\phi=-\pi/2}^{\phi=+\pi/2} \dfrac{cos\varphi Rd\phi dh}{[(rcos\Theta - Rcos\phi)^2 + (rsin\Theta - Rsin\phi)^2 + (z-h)^2]}$	(7.9)
Volume Source	----	$E_\lambda = \dfrac{k_{\lambda 3}}{4\pi} s \displaystyle\int_{h=-L}^{h=+L} \displaystyle\int_{\eta=0}^{\eta=R} \displaystyle\int_{\phi=-\pi}^{\phi=+\pi} \dfrac{\eta d\eta d\phi dh}{[(rcos\Theta - \eta cos\phi)^2 + (rsin\Theta - \eta sin\phi)^2 + (z-h)^2]}$	(7.10)

Source: Boyjoo, Y., Ang, M., Pareek, V., 2013. Copyright Elsevier.

Note:

E_λ = incident intensity at particular wavelength

$k_{\lambda 1}$ = rate of photons emitted per unit length of lamp

$k_{\lambda 2}$ = rate of photons emitted per unit surface of lamp

$k_{\lambda 3}$ = rate of photons emitted per unit volume of lamp

h = axial coordinate of an element on lamp

z = distance.

7.3.3   STEP III: RADIATION MODELING

The solution of radiation transport equations (RTE) is a necessary part in modelingg of photocatalytic reactors. RTE is related with energy loss due to absorption, out-scattering, and energy gain because of in-scattering of photons [7–9]. The RTE equations related with radiation modeling for slurry and immobilized systems are given in Table 7.2:

TABLE 7.2
Modeling Equations for Slurry and Immobilized Reaction Systems [7]

Slurry Systems

$$\frac{dI_\lambda(S,\Omega)}{dS} = -\kappa_\lambda I_\lambda(S,\Omega) - \sigma_\lambda I_\lambda(S,\Omega) + \frac{1}{4}\sigma_\lambda \int_0^{4\pi} p(\Omega' \to \Omega) I_\lambda(S,\Omega') d\Omega' \qquad (7.11)$$

$$E_\lambda(S) = \int_{\Omega=0}^{\Omega=4\pi} I_\lambda(S,\Omega) d\Omega \qquad (7.12)$$

$$L_\lambda^a(s) = \kappa_\lambda(S) E_\lambda(S) \qquad (7.13)$$

$$L^a(S) = \int_{\lambda min}^{\lambda max} \kappa_\lambda(S) E_\lambda(S) \qquad (7.14)$$

$$\langle L^a \rangle = \frac{1}{V_W} \int_{V_W} L^a(S) dV_W(S) \qquad (7.15)$$

$$\phi_{overall} \frac{\langle r_p \rangle}{\langle L_p^a \rangle} \qquad (7.16)$$

Immobilized systems

$$I_O = I_m - I_{ref} \qquad (7.17)$$

$$\frac{dI_{0,z}}{dz} = -\alpha_\lambda I_{0,z} \qquad (7.18)$$

$$\frac{dI_\lambda(S,\Omega)}{dS} = -\sigma_\lambda I_\lambda(S,©) + \frac{1}{4\pi}\sigma_\lambda \int_0^{4\pi} p(\Omega' \to \Omega) I_\lambda(S,\Omega') d\Omega' \qquad (7.19)$$

Note:

S = rate of mass transfer, Ω = solid angle,
λ = wavelength, σ = scattering coefficient,
α = volume fraction, V = volume,
W = normal velocity, I = intensity,
Z = distance, ϕ = quantum yield,
K = absorption coefficient, $p(\Omega - \Omega')$ = phase function for in-scattering of radiation,
E = incident intensity, p = pressure shared by all phases,
L^a = local volumetric rate of energy absorption.

7.3.4   STEP IV: KINETIC MODELING

In the slurry photocatalytic reactor, kinetic modeling helps to present a path for the degradation of contaminants in the wastewater [10, 11]. This model considers all steps which are accountable for photocatalytic reactions in several processes like adsorption, electron-trapping, catalyst activation, electron-hole recombination, etc. The rate model equations for slurry and immobilized systems are given as below.

7.3.4.1   Slurry System Rate

Photocatalytic reaction depends upon the rate of degradation of a compound in the suspension of aqueous TiO_2. Except this, other important dependent factors, such as operating parameters, are the concentrations of oxygen, TiO_2, and pollutant, temperature, pH, and radiation [11]. Therefore, kinetic modeling expressions are as follows:

$$-(\text{rate}) = k.f([O]) \tag{7.20}$$

$$k = f(temperature,\ O_2\ concentration,\ W_{cat},\ pH,\ E,\ radiant\ flux) \tag{7.21}$$

where
[O] = pollutant concentration
E = intensity of incident radiation at any point from all the directions
W_{cat} = catalyst loading
k = rate constant.

The intensity of the radiation can be modified by placing metal screens between the lamp and reactor.

$$-(\text{rate})_R = kE^m f([O]) \tag{7.22}$$

Distance can be managed by reflection of incident light radiation within a slurry system. The influence of this light intensity can be given with the following rate expression:

$$-(\text{rate})_R = k\langle L^a \rangle L^a f([O]) \tag{7.23}$$

where
m = mass
$\langle L^a \rangle$ = local volumetric rate of energy absorption.

For slurry reactors with large annulus, the rate equation can be written as

$$-(\text{rate})_R = k_1(f([O])[\langle L^a \rangle - K_2 W_{cat}] \tag{7.24}$$

7.3.4.2   Immobilized System Rate

The coating of immobilized systems is done with thin layer of TiO_2. The estimation of accurate absorption of radiation is not an easy task in these kinds of systems. Therefore, the accurate rate equation can be defined as Eq. (7.22) by assuming the ratio of incident radiation to absorbed radiation is equal to unity for a given catalyst in this photocatalytic reaction system.

7.3.5   BOUNDARY CONDITIONS

Initial conditions to model a reactor are recognized as boundary conditions to describe the simulation procedure for hydrodynamic, radiation, and kinetic modeling of the photocatalytic reactor as well [7]. The details are as follows:

7.3.5.1   Hydrodynamic Modeling-Boundary Conditions

(a) Slurry systems-gas distribution modeling

A uniform surface is required for the modeling of the gas inlet for a slurry reaction system as compared to individual holes for achieving good approximation for further computation.

(b) Slurry system-gas outlet

The gas outlet section should be long enough to avoid the modeling of the gas disengagement region so that the top surface of the liquid is modeled to set the normal gas and liquid velocity to zero. The cells are attached to the top of the surface of the liquid and are known as the sink to escape the formation of bubbles. This sink can be mathematically represented by the following expression:

$$S_G = -A_B \alpha_{GB} W_{GB} \rho_G \qquad (7.25)$$

where
A_B = area of the bottom surface of a cell attached to the top surface
α_{GB} = normal velocity of gas bubble
W_{GB} = the gas volume fraction of the cell attached to top surface
ρ_G = gas density.

7.3.5.2   Radiation Modeling-Boundary Conditions

After hydrodynamic modeling, the radiation modeling-boundary conditions component wise will be applied as follows:

(a) Lamp surface

The modeling equations are represented in equations from Eqs. 7.6 to 7.10 in earlier sections. The UV detectors are set up next to the lamp surface to get better results according to the rate of photon emission for specified wavelengths.

(b) Reactor wall

As we know, some of the incident radiation on the wall is absorbed, reflected, and the rest is transmitted.

For the transparent wall,

$$a + t + r_{ref} = 1 \qquad (7.26)$$

where
a = absorptivity on reactor wall surface,
t = transmittance of reactor wall,
r_{ref} = reflectivity of reactor wall.

For a blackbody wall, after applying Kirchoff's law

$$a = e \qquad (7.27)$$

If the reactor wall is non-transparent,

$$t = 0 \qquad (7.28)$$

Therefore, after applying the conditions from Eqs. (7.27) and (7.28), Eq. (7.26) can be rewritten as

$$e + r_{ref} = 1 \qquad (7.29)$$

The values of reflectivity as per requirement can be obtained from the literature.

The fouling is a common issue in transparent walls of slurry reactors because the particles of the TiO_2 photocatalyst are deposited on the surface of the reactor wall. To encounter this fouling issue, a fouling coefficient was introduced as discussed in the following equations:

$$T_{total} = t_w t_{fou} \qquad (7.30)$$

where
t_w = wall transmittance
t_{fou} = transmittance due to fouling = $Y_{fou} W_{cat}$ $\qquad (7.31)$
Y_{fou} = fouling coefficient (m^3g^{-1})

(c) Reactor window

The emission source is separated from reaction medium via the reactor window. Due to reflection, the incident radiation flux at the reactor window is expressed from following relation:

$$E_{in}^{window} = \int I_{in} s.\hat{n} d\Omega \qquad (7.32)$$

where,

s = the direction of the incident radiation,

$\hat{n}$ = unit vector normal to the plane of incidence.

The net radiative flux leaving the reactor window is given as below:

$$E_{out}^{window} = r_{windows} E_{in}^{window} + E_{transmitted}^{window} \qquad (7.33)$$

where,

r_{window} = reactor window reflectivity,

$E_{transmitted}^{window}$ = transmitted portion of radiation.

7.4 EFFECT OF PARAMETERS

There are many operating parameters, which affect the efficiency of any photocatalytic reactor. Some are discussed in detail below:

7.4.1 PHOTOCATALYST COMPOSITION AND TYPE

Activity of a photocatalytic semiconductor is dependent on its structure and surface particles, such as particle size, porosity, surface area, crystal composition, band gap, and surface hydroxyl density. The efficiency of a heterogeneous catalyst depends on particle size and surface area [12]. Degradation of several catalyst compounds with some organic compounds in liquid solution has been examined thoroughly. Different compounds such as Degussa P-25, Hombikat, PC10, UV100, PC50, PC500, Rhodia and Travancore Titanimum Product (TTP) have also been examined. P-25 consists of 25% rutile and 75% anatase having a particle size 20 nm and BET surface area of 50 m²/gm. Hombikat UV100 was fabricated from pure (~100%) anatase with particle size 20 nm and BET surface area of 250 m²/gm. While PC500 was fabricated from pure (~100%) anatase with a particle size between 5 to 10 nm and BET surface area 287 m²/gm. During comparison of Degussa P-25, Hombikat, and PC500, it was observed that the degradation rate was found highest in the presence of P-25. Efficacy of different photocatalyst is generally ranked as follows: UV100 > PC500 > TTP. The photocatalytic activity of P-25 was explained by its mixed composition and crystalline anatase and rutile. Rutile has negative reduction power and plays an important role inhibiting the electron-hole restructurization in anatase. Anatase has several transition points between two phases that enhance the electron carrier and simultaneously enhance the photo catalyst activity.

BET surface area, lattice mismatches, density of hydroxyl groups, and other impurities affect the absorption behavior of pollutants. Other factors such as the recombination rate of electronic hole pairs was subsequently leading to the variation in photo catalytic activities.

7.4.2 Catalyst Loading

Pollutant degradation rate depends on the catalyst loading. Initial degradation first increases and thereafter decreases with the enhancement of catalyst loading due to the screening effect of light. When reaction rate goes to infinity level, it means that it is independent of the concentration of titanium dioxide. This effect reduces the photocatalyst surface area and photocatalyst activity. Optimum dose of the photocatalyst is necessary for obtaining the photocatalytic degradation efficiency [12,13].

Some researchers have reported the effect of photocatalyst degradation on the catalyst loading due to difference in photon wave length, light intensity, radiation fluxes, and working geometry. For an application of industrial wastewater treatment, the concentration of titaniumdioxide photocatalyst influences directly the entire system.

7.4.3 Effect of pH

pH is the other affecting operating parameter in the heterogeneous photocatalytic process. The variation in pH affects the catalyst surface charge, particle size, and the condition of valance band and conduction band. Organic pollutants are different in terms of hydrophobicity and water solubility [7,12,13]. In such circumstances, some compounds remain unchanged (manmade waste or natural waste) and certain pH can distinguish the compounds in terms of specification and physicochemical properties. Variation of pH leads to the variation in the isoelectric point or surface charge of the photocatalyst used. Value of pH at which charge is neutralized is known as the point of zero charge. The photocatalytic compound titaniumdioxide has point zero charge between 4.5 and 7.0, which depends on the composition. The value of PZC is minimum in the absence of electrostatic force due to the interaction between water contaminants and photocatlyst particle. The photocatalyst surface is positively charged and it exerts an electrostatic attraction force towards negative charge compounds, if the operating pH value is less the point zero charge. pH value is more than PZC, if the photocatalyst surface is negatively charged with subsequent exorcising anionic compound. Under acidic condition, the surface ionization state of the photocatalyst can be protonated and under alkali condition it is deprotonated. The water equilibrium conditions are given below:

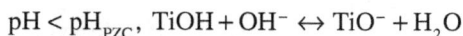

$$pH < pH_{PZC}, \ TiOH + H^+ \leftrightarrow TiOH_2$$

$$pH < pH_{PZC}, \ TiOH + OH^- \leftrightarrow TiO^- + H_2O$$

The value of point zero charge of titaniumdioxide is 6.25. At pH < 6.25, the titaniumdioxide surface is positively charged in acidic medium and is negatively charged during alkaline condition (pH > 6.5). Increase in pH value of the solution can be seen as the effect of hydroxyl radicals.

Reaction is observed between titaniumdioxide-induced holes and hydroxide ions of the catalyst. At oxidation phase, the low value of pH is important as compared to the neutral and high levels of pH. Presence of more hydroxyl ions generates higher hydroxyl radicals at the surface of titaniumdioxide. At high value of pH, the degradation efficiency will rise logically. Several researchers have also been observed that the degradation of organic compound was found effective on titaniumdioxide surface with the variation of pH value. Electrostatic repulsion and attraction between the organic molecule in ionic form and the surface of the photocatalyst can be varied with change of pH value. Hence, pH is an important parameter that can be controlled variable for the enhancement of organic pollutants' degradation. Different pollutants have different optimum pH values under optimal photocatalytic degradation. Each pollutant's pK_a value is used to predict the catalytic activity at particular pH value because this affects the absorption property of the catalyst and stability of pollutant molecules. Acidic value of the pollutant solution can be estimated by the pK_a value. pK_a value is inversely proportional to the strength of acidity. At weaker strength, the value of pK_a is higher. A suitable pH control strategy is required in wastewater treatment through photocatalytic reaction.

7.4.4 POLLUTANT CONCENTRATION AND TYPE

During wastewater treatment, there is a very important relation between substrate concentration and photocatalytic degradation rate. With increase in pollutant substrate concentration, the probability of collision between pollutant and the surface of a photocatalyst is higher. Consequently, the degradation rate is higher due to short life time of hydroxyl radicals responsible in oxidizing the organic pollutants [7,12,13]. Due to enhancement the substance concentration upto a certain limit, the intermediate species will be generated and absorbed on the photocatalyst surface. Since the photodegradation process is the non-selective process and thus the intermediates are obtained due to decomposition of the pollutant, which produces the superoxide and hydroxide radicals. As a result, the photocatalyst active sites are compromised. Hence, absorption rate is slower due to overall low degradation rate. It can be deduced at low substrate concentration, rate of degradation is less dependent on the number of active sites of the catalyst. During first-order kinetics, it is proportional to the substrate concentration. Another factor is the molecular structure of the pollutant, which affects its degradation extent.

The Langmuir–Hinshelwood (LH) model explains the kinetics of the heterogeneous catalytic process. The following expression is given to explain the kinetic of the heterogeneous catalytic system.

$$\frac{1}{r} = \frac{1}{-\left\langle \dfrac{dC}{dt} \right\rangle} = \frac{1+KC}{k_r KC} \qquad (7.34)$$

In Eq. (7.34), r represents the rate of reaction, which changes with time.

At initial rate of reaction

$$\frac{1}{r_o} = \frac{1}{-\left\langle \dfrac{dC}{dt} \right\rangle} = \frac{1 + KC_o}{k_r KC_o} \qquad (7.35)$$

and

$$\frac{1}{r_o} = \frac{1}{-\left\langle \dfrac{dC}{dt} \right\rangle} = \frac{1 + KC_e}{k_r KC_e} \qquad (7.36)$$

where
r_o is the initial reaction rate, C_o is the initial dye concentration and C_e is the dye concentration at equilibrium obtained after completing the dark experiment. Parameters in Eqs. (7.35) and (7.36), k_r and K_o are functions of C_o and C_e, which can be linearized as

$$\frac{1}{r_o} = \frac{1}{k_r} + \frac{1}{k_r KC_o} \qquad (7.37)$$

where
$KC \ll 1$ for first-order kinetics.

Recently, some literature showed $KC \ll 1$ was estimated at zero-order kinetics in the LH model. However, LH expression is written at $KC_o \gg 1$ and the LH kinetics will not be reduced for zero-order kinetics.

7.4.5 LIGHT INTENSITY AND WAVE LENGTH

Light intensity is another source that initiates heterogeneous photocatalytic and photo-chemical processes. During photochemical process, electron-hole pair formation and rate of initiation of photocatalysis depend on the intensity of light. Distribution of light in a reactor indicates the pollutant degradation rate and the pollutant conversion efficiency [13]. Pollutant degradation rate behaves linearly as the intensity of light increases. But, at high intensity, the pollutant reaction rate will be independent of the intensity of light. Reaction involving the formation of an electron hole is more significant than that of recombination of involving electron hole at low intensity. With enhancement of light intensity, the electron-hole formation rate at the surface of titanium dioxide rises and consequently also increases the ability to decompose organic pollutant. At particular cases, the rate constant of first order of iodosulfuran degradation was found directly proportional to the photon flux, when the flux value is below to 10^{16} photon $1/s.cm^2$. The oxidizable organic substrate concentration is higher than the rate of formation at electron-hole pairs within the photocatalyst at low flux value. This exhibits a linear relationship between photo flux and rate of degradation. At elevated photon flux, the value of the rate constant (10^{16} photon 1/s.cm^2) varies with the square root of the flux, in which the rate of degradation experiences immense reduction due to the recombination of electron-hole pair.

Some researchers have claimed the titaniumdioxide photocatalytic degradation rate is not influenced by the intensity of light of little photon energy (low 1μW cm^{-1}). To obtain a high photocatalytic reaction rate in the application of wastewater treatment, high light intensity is required to activate the sites of titaniumdioxide. However, the application of titaniumdioxide photocatalyst reaction is limited to photon at a shorter wave length of 400 nm such as UV light. At wave length < 400 nm, the degradation rate is linear with incident radiant flux. It has also been observed with polychlorinated biphenyls and texim destruction with the presence of high intensity of photons. A good linear correlation was found between the light intensity of first-order rate constant. In some cases, photocatalytic oxidation rate is also dependent on the intensity of light. Therefore, in preliminary studies of microbial consortia presence and photoreactor performance in wastewater application are used to estimate the minimum irradiation at irradiance constant.

7.4.6 TEMPERATURE EFFECT

Several experimental studies have been performed on the impact of temperature at the degradation of the pollutant rate as well as the photocatalytic reaction [12,13]. A sufficient amount of energy is required to activate the surface of titaniumdioxide under the presence of illumination of natural sunlight. Its perception was extrapolated for such dependency. The photocatalytic reaction temperature of more than 80°C promotes charge-carrier recombination and thus inhibits the organic compound on the surface of titaniumdioxide. In most of the cases, the photocatalytic systems are operated at room temperature, which means no more heating is required. Generally, the recommended temperature range for the photodecomposition process is 20 to 80°C. Desorption rate of the final product was at 0°C. Due to this reason, the apparent activation energy increases. At temperature above 80°C, the absorption factor for the reactant will be limited and resultant activation energy will be negative. In the case of photodecomposition, the rate increases with the increase in reaction temperature within the range of 20°C to 60°C.

7.5 CONCLUSIONS

The present chapter reviewed and compiled important information regarding the photocatalytic reactor in terms of its characteristics, modeling using chemical fluid dynamics with boundary conditions, and the effects of various operational parameters. The summary of each section is given below:

(i) The types of photocatalytic reactors, such as immobilized and suspended form, are discussed with specified requirement to carryout photocatalytic reaction.

(ii) The specification for the characterization of the photocatalytic reactor along with several essential factors, such as shape, type of photocatalyst, type of phases, and others.

(iii) The step-by-step formulation for modeling the photocatalytic reactor has been explained in detail by using chemical fluid dynamics in which various types of models can be used for a variety of applications. For large-scale systems, E–E model is preferred over the E–L model due to its specific computational requirements whereas the E–L model is used for low catalyst concentration.

The linear source model provides a realistic approach for surface and volume source models using chemical fluid dynamics. Furthermore, kinetic modeling provides a clear picture for contaminants'degradation to conduct experiment to see photocatalytic activities, such as catalyst deactivation, adsorption, hole trapping, electron trapping, and electron-hole recombination, etc.

(iv) The effects of various operational parameters (such as temperature, sunlight intensity, wavelength of solar spectrum, types of pollutants, types of catalysts, catalyst loading, pH, pollutant concentration, etc.) on photocatalytic activity has been discussed in detail in the present chapter.

At last, the present chapter overviews the hydrodynamics, radiation, and kinetic modeling of the photocatalytic reactor using the formulation of chemical fluid dynamics for the degradation of a variety of contaminates present in wastewater streams with the help of renewable energy, i.e., solar energy at optimum scale.

REFERENCES

[1] Kumar, B., Kumar, S., Kumar, S., 2018. Butanol reforming: an overview on recent developments and future aspects. Reviews in Chemical Engineering 34:1. https://doi.org/10.1515/revce-2016-0045

[2] Kumar, B, Singh, L., Rekha, P., Kumar, P., 2022. Photocatalyticbiomass valorization into valuable chemicals (chapter 1), Handbook of Biomass Valorization for Industrial Applications, Edited by Shahid-ul-Islam, Aabid Hussain Shalla, Salman Ahmad Khan, First Edition, Wiley Online Library, Pages: 1–12. https://doi.org/10.1002/9781119818816.ch1

[3] Tong, K., Yang, L., Du, X., 2020. Modelling of TiO_2-based packing bed photocatalytic reactor with Raschig rings for phenol degradation by coupled CFD andDEM, Chemical Engineering Journal 400:125988. https://doi.org/10.1016/j.cej.2020.125988

[4] Assadi, H., Armaghan, F., Taher, R.A., 2021. Photocatalytic oxidation of ketone group volatile organic compounds in an intensified fluidized bed reactor using nano-TiO_2/UV process: An experimental and modeling study, Chemical Engineering and processing–Process Intensification 161:108312. https://doi.org/10.1016/j.cep.2021.108312

[5] Visan, A., Ommen, J.R., Kreutzer, M.T., LammertinkR.G.H., 2019. Photocatalyticreactor design: Guidelines for kinetic investigation, Ind. Eng. Chem. Res. 58:5349. https://doi.org/10.1021/acs.iecr.9b00381

[6] Wang, D., Mueses, M.A., Mˊarquez, J.A.C., Machuca-Martínez, F., Grˇciˊc, I., Moreira, R.P.M., Puma, G.L., 2021. Engineering and modeling perspectives on photocatalytic reactors for water treatment, Water Research 202:117421. https://doi.org/10.1016/j.watres.2021.117421

[7] Boyjoo, Y., Ang, M., Pareek, V., 2013. Some aspects of photocatalytic reactor modeling using computational fluid dynamics, Chemical EngineeringScience 101:764. https://doi.org/10.1016/j.ces.2013.06.035

[8] JanczarekM., Kowalska, E., 2021. Computer Simulations of Photocatalytic Reactors, Catalysts 11:198. https://doi.org/10.3390/catal11020198

[9] Ballari, M.M., Satuf, M.L., Alfano, O.M., 2019. Photocatalytic Reactor Modeling: Application to Advanced Oxidation Processes for Chemical Pollution Abatement, Topics in Current Chemistry 377:22. https://doi.org/10.1007/s41 061-019-0247-2

[10] Malayeri, M., Haghighat, F., Le C.S., 2021. Kinetic modeling of the photocatalytic degradation of methyl ethyl ketone in air for a continuous-flow reactor, Chemical Engineering Journal 404: 126602. https://doi.org/10.1016/j.cej.2020.126602

[11] Malayeri, M., Lee, C.S., Niu, J., Zhu, J., Haghighat, F., 2021. Kinetic and reaction mechanism of generated by-products in a photocatalytic oxidation reactor: Model development and validation, Journal of Hazardous Materials,419:126411. https://doi.org/10.1016/j.jhazmat.2021.126411

[12] Roushenas, P., Ong, Z.C., Ismail, Z., Majidnia, Z., Ang, B.C., Asadsangabifard, M., Onn, C.C., Tam, J.H., 2018. Operational parameters effects on photocatalytic reactors of wastewater pollutant: A review, Desalination and WaterTreatment 120:109. https://www.deswater.com/DWT_abstracts/vol_120/120_2018_109.pdf

[13] Reza,K.M., KurnyA.S.W., GulshanF., 2015. Parameters affecting the photocatalytic degradation of dyes using TiO_2: A review, Applied Water Science 7:1569. https://doi.org/10.1007/s13201-015-0367-y

8 Computational Plasma Dynamics for Wastewater Treatment Systems

Anbarasan Rajan and
*Mahendran Radhakrishnan**
Centre of Excellence in Non-Thermal Processing,
National Institute of Food Technology, Entrepreneurship
and Management (NIFTEM)–Thanjavur, India
(an Institute of National Importance; formerly Indian
Institute of Food Processing Technology–IIFPT), Thanjavur,
Tamil Nadu, India.
*Corresponding Author

CONTENTS

DOI: 10.1201/9781003325147-8

8.1   INTRODUCTION

Water is an essential natural resource for humans, animals, and most other life forms. Due to industrialization and urbanization, water resources are being contaminated with pollutants, such as pesticides, dyes, plastics, pathogenic microbes, oil, synthetic fertilizers, and chemical compounds like detergents, pharmaceuticals (Saravanan et al., 2021; Vörösmarty et al., 2010). With the increasing world population, the quantity of wastewater released from industrial, agricultural, and domestic sectors are also increasing day by day; which in turn causes freshwater scarcity and leads to the unfair distribution of water among these major sectors (Ezugbe & Rathilal, 2020). In addition, these pollutants might transfer through the aquatic systems to humans and can cause health effects (Schwarzenbach et al., 2010). Hence, researchers were pushed to develop various physical, chemical, and biological wastewater treatment methods in the recent past (Zeghioud et al., 2020). Nevertheless, the method selection and processing condition play a major role on the efficiency of treatment as the effluent nature varies based on the origin (Crini & Lichtfouse, 2019). However, the treatment plant effluent is itself a source of pollutant as it contains trace organic matters and cannot be removed completely by physical and biological methods (Miklos et al., 2018). Though the chemical methods can remove the organic pollutants, they also pose a threat to the eco-system through the formation of inorganic solid and unreacted chemical residues. Hence, researchers are focusing on alternative methods to solve the pollutant issue, one such method is cold plasma (CP). It is widely used among different sectors due to its simplicity and eco-friendly nature. Thus, it has also started to gain more attention in wastewater treatment as it can degrade various chemical and biological pollutants (Zeghioud et al., 2020).

Plasma consists of photons, electrons, ions, atoms, free radicals, excited and non-excited molecules at different concentrations based on the energy given to the system. In general, these components can be categorized as reactive species (RNS, reactive nitrogen species; ROS, reactive oxygen species), UV photons, and accelerated electrons/atoms. These components are good oxidizers of chemicals and capable of causing DNA damages to the living cells. Hence, both reactive species (RS) and UV photon are responsible for effectiveness of wastewater treatment, however, the treatment efficiency may vary based on the supplied voltage, gas-composition, gas-flow, barrier property, and electrode distance (Mahendran & Alagusundaram, 2015; Potluri et al., 2018). In addition to plasma components, the system configuration and the type of plasma used for treatment play a major role on deciding the treatment effectiveness. Basically, plasma type can be categorized into thermal and non-thermal based on the thermal equilibrium state of its constituents. However, based on the type of discharge, the classification can be further expanded as follows (Jiang et al., 2014):

   a.   Pulsed corona/streamer/spark discharge
   b.   DC pulseless corona discharge
   c.   Dielectric barrier discharge
   d.   Gliding arc discharge
   e.   DC glow discharge
   f.   DC arc discharge

Nevertheless, it is also important to know the zone of plasma interaction since the wastewater and plasma reaction (contact) interface highly influence the treatment irrespective of the other parameters like system configuration, electrode distance, treatment time, and gas composition (Ranjitha Gracy et al., 2019; Zeghioud et al., 2020).

Thus, each parameter plays its part during the plasma treatment, it is very tough to optimize the process condition during treatment; meanwhile, performing multiple trials for the treatment optimization involves both time and energy (fuel, cost, labor). However, CFD modeling enables a way to predict and optimize plasma treatment by utilizing appropriate physics and operational parameters as inputs to the model. In general, CFD is a simulation tool that models fluid flow behaviors using high-end computers and applied mathematics for process and design optimization. This technique involves solving equations for the conservation of mass, momentum, and energy; and subsequently utilizes numerical methods for forecasting velocity, temperature, and pressure distribution within the system. Once the results are computed, the model provides three-dimensional images for the visualization of fluid behaviors (Anandharamakrishnan, 2013). However, based on the process complexity, hardware support, and available computation time, the simulation can be made with either 1D, 2D, or 3D geometry (space dimension). The major advantage of using CFD techniques for process modeling is that they provide both stationary and time-dependent outputs for any given condition with high accuracy, which ultimately reduces the unwanted real-time experiments (Ersion, 2013; Shen et al., 2020). Thus, it is possible to model both stationary and unsteady state plasma process and extract essential information using CFD techniques for wastewater treatment. These data enable the precise optimization of wastewater remediation process for any given plasma operating condition and also leads to the forecasting of treatment output for the fixed process parameters. Since the plasma modeling is performed through CFD technique this modeling process is called as computational plasma dynamics (CPD). In this chapter we discuss the existing wastewater treatment studies to understand the importance of plasma operational condition on the pollutant removal. With that note, we also discuss the CPD modeling approach for wastewater treatment and interpretation of the model results with the actual process conditions.

8.2 EFFECT OF COLD PLASMA PROCESSING CONDITIONS ON WASTEWATER TREATMENT

The oxidative components present in plasma causes two distinguishing effects in the water purification process. The first one resulting from direct electron collisions while the other one is due to ionic, molecular (O_3, H_2O_2, O_2), pyrolysis and photolysis products ($O\bullet$, $O_2\bullet^-$, $\bullet OH$, and $H\bullet$) reactions. Apart from that, the pollutants present in water can also be destroyed by aqueous electron and $H\bullet$. When it comes to microbial inactivation, plasma discharge causes physical and chemical reactions, which leads to cell membrane damage, cell structure deformation, and UV-induced mutation. The plasma discharge can be of any of the following four ways (Fig. 8.1): (a) plasma discharge above the water surface, (b) direct discharge in the liquid phase,

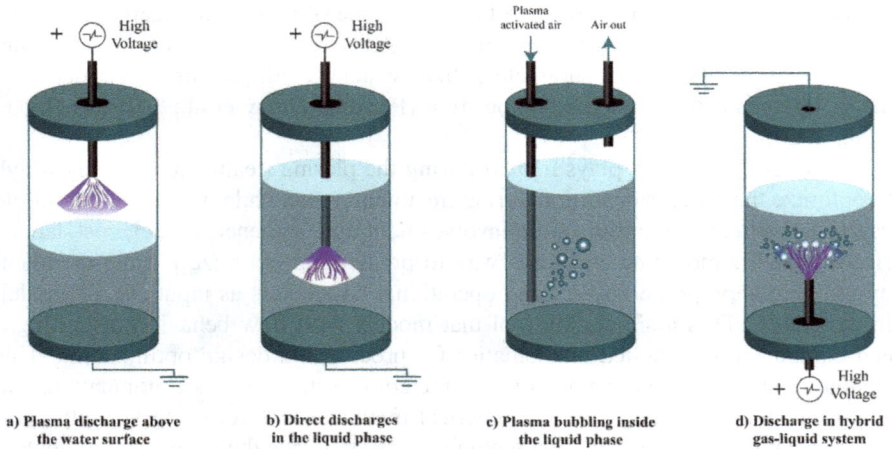

FIGURE 8.1   Different type of plasma discharge for polluted wastewater treatment. (a) Plasma discharge above the water surface. (b) Direct discharge in the liquid phase. (c) Plasma bubbling inside the liquid phase. (d) Discharge in hybrid gas-liquid system.

(c) plasma bubbling inside the liquid phase, and (d) discharge in hybrid gas-liquid system (Ranjitha Gracy et al., 2019; Zeghioud et al., 2020).

The efficiency of plasma treatment on wastewater treatment varies based on different parameters such as time, voltage, electrode distance, pulse repetition rate, gas composition, and air flow rate. Meropoulis et al. (2021) used five different dye solutions (SY – Sunset Yellow, OII – Orange II, MO – Methyl Orange, MB – Methylene Blue, MV – Methyl Violet) and treated them using nanosecond pulsed (NSP) dielectric barrier discharge (DBD) plasma to investigate the effect of plasma operating conditions. It was observed 20.5%, 39.8%, and 50.7% degradation in the OII dye after exposure of 2 min at 23.8 kV, 27.0 kV, and 31.4 kV, respectively. In addition, further increase of treatment time showed nearly complete removal of OII dye at 27.0 kV and 31.4 kV voltage levels (88.8% in 23.8 kV). Apart from supplied voltage and treatment time, other parameters such as flow rate and pulse frequency also play an important role in degradation efficiency. Therefore, the author used air flow rate in the range of 0.1 to 1.0 liter/min and observed the maximum degradation of dyes at 0.2 liter/min flow rate. However, when the air flow rate was increased above 0.5 liter/min, the dye degradation started to deplete. It was stated that the reduction in degradation was due to the limited retention time of air inside the treatment system that reduced the interaction of plasma air with the liquid surface. Hence, the transport of reactive species into the water dye mix was minimal during the treatment.

In terms of frequency, the degradation was positively influenced by higher plasma frequency ranges. Nevertheless, the treatment effectiveness was not the same for all the dyes and it changed based on the dye types and its concentrations. In particular, SY dye showed more resistance to plasma treatment than any other dyes because of its complexity in chemical structural. Similar studies carried out by others researchers on dye reduction are shown in Table 8.1.

Since the major principle of plasma-assisted wastewater treatment is the reactive species (RS)-induced pollutant's oxidation, it is important to understand the characteristics of plasma RS and its influence on the treatment effectiveness. Wang et al. (2017) treated MB dye (100 mg/L) in a double-chamber DBD reactor with two different (air and O_2) carrier gases with fixed flow rate (60 std. cm^3/min) and input voltage (6 kV). As the result, a complete degradation was achieved in O_2 plasma treatment MB samples when the air plasma treatment struggled to degrade even 85% of the dye after 100 min. Thus, it was identified that the MB dye reduction is largely associated with ROS (NO_2^-, NO_3^-, O_3, and H_2O_2) than the RNS as the RNS have low oxidizing potential than ROS. Furthermore, the major drawback of air plasma is that they produce more RNS that consumes a large amount of O_2 and leave very few O_2 molecules to form ROS species. A similar study carried out by Sun et al. (2012) also reported a comparable result where the reaction rate constant of O_2, Air, N_2, and Ar gases were found be 1.38×10^{-1}/min, 1.14×10^{-1}/min, 2.51×10^{-2}/min, and 3.09×10^{-2}/min, respectively, for MO dye. In addition, the experiment was also performed by varying the electrodes (needles and ground) distance from the water surface to understand its effect on MO degradation. When ground electrode was kept away from the water surface, more discharge current passed through the MO solution thus it increased the degradation up to ~85% (at 28 mm away from water surface). Meanwhile, reducing the distance between the needles and water surface was found to be effective for plasma treatment as it reduced more than 84% colorant when the distance was fixed at 1 mm. However, the most critical finding of this experiment was the sample's conductivity-dependent MO degradation. It was found that the conductivity of 1700 µs/cm was suited for higher efficiency of plasma treatment as it showed good degradation for all exposure times. However, further increment (cause reduction in discharge spots) or reduction (low discharge through sample) in the conductivity reduced the degradation efficiency of the plasma treatment.

Another major source of water pollution is the pharmaceutical wastes, which can be degraded using non-thermal plasma treatment. Magureanu et al. (2021) discussed elaborately pharmaceutical wastewater contamination in their review. Antibiotics such as cefixime also contaminate water and it can be effectively treated using plasma. Zhang et al. (2021) used plasma bubbling for degrading cefixime and achieved up to 99.8% degradation when using O_2 as the carrier gas. In this study, the author observed an increase in the degradation of cefixime while increasing the sample pH and this effect was clearly evident after 5 min of treatment where the sample with pH 10 showed up to 75.3% antibiotics reduction, while the other one with a pH of less than 5 had not even caused 50% reduction. The study also revealed the negative impact of increased antibiotics' concentration on plasma treatment efficiency. A similar pH-induced antibiotic degradation trend was observed in Rong & Sun's (2014) study, however, a pH of more than 11 caused reduction in the sulfadiazine degradation. The OH species generated at high pH are prone to react with carbonate ions and due to which the sulfadiazine degradation reduced in the higher pH range (similar studies on pharmaceutical/antibiotic degradation are shown in Table. 8.1). Magureanu et al. (2021) have given a detailed review on antibiotics' removal from wastewater using non-thermal plasma. It is also noteworthy that the over use of antibiotics causes resistance in microorganisms and thus, increases the chance of microbial contamination. However, plasma treatment is capable of reducing antibiotics as well as microorganisms. Liao et al. (2021) used ACP at 0.71 kJ/cm^2 intensity and found up to 99.999% reduction in the antibiotics-resistant *Escherichia coli* population due to cell membrane depolarization. Similar results were obtained in Patinglag et al.'s (2021) study on microfluidic reactor assisted wastewater treatment. After plasma treatment samples were observed in SEM where *Pseudomonas Aeruginosa* and *E. coli* were found do have damages on the cell structure while the untreated cells remained intact. The cell damage caused by microfluidic the plasma reactor was found to be sufficient enough to cause complete cell death after 5 min of treatment.

Another aspect of wastewater treatment is that the secondary reactive species generation inside the sample and their impact on degradation process. Meiyazhagan et al. (2020) studied the textile dye degradation and evaluated the amount of H_2O_2 and OH radical production in contaminated water during the plasma treatment. It was found that the use of scavengers (for OH radical) in the dye solution (1% and 2% dimethyl sulfoxide) reduced the degradation efficiency (< 80%) of the plasma treatment as compared to the other treatment (~100%) that was carried out without any scavengers. So the free radicals (i.e., OH, H_2O_2) produced inside the water plays the major role on pollutants' degradation, hence it is important to quantify the reactive species generation inside the liquid systems along with the RS produced in the air space.

From the above research findings and Table. 8.1, it is clear that both process conditions and sample characteristics influence the treatment efficiency. However, it is not always possible to experiment all the process and product variable combinations for pollutant (dye, pesticide, and antibiotics) degradation studies as it consumes both time and energy. Thus, tresearchers started focusing on CFD modeling approaches to simulate treatment conditions to predict the characteristics of generated plasma and their degradation efficiency through different equations and chemical reactions.

8.3 COMPUTATIONAL PLASMA DYNAMICS

Cold plasma is the fourth state of matter in which the gas at quasi-neutral ionized state will be having charged particles (electrons and ions), atoms, ground level or excited state molecules. Among these species, the free electrons alone will be having the temperature above 10,000 K in the plasma state while the heavy species temperature only varies between 300–1000 K. Due to the species temperature difference, cold plasma is being referred to as non-thermal plasma (Pankaj et al., 2018). When it comes to species movement, the transport of heavier species is much less than the other species. Hence, the electrons and other species interact (react or collide) between/within them and also with the plasma system walls (Manoharan & Radhakrishnan, 2021). Cold plasma also produces photon when the exited particles generated inside the reactor comes to the ground stage; thus, the cold plasma generation process is a merge of fluid mechanics, reaction engineering, physical kinetics, heat transfer, mass transfer, and electromagnetics (Ersion, 2013).

Due to the complexity and poor understanding of physical and chemical processes involved in the plasma generation process, it is difficult to control, reproduce, and optimize plasma with the desired properties (i.e., reactive species concentration). Thus, it

TABLE 8.1
Plasma Technology Applications in Wastewater Treatment

Sample	Treatment Condition	Experimental Results	Reference
Diluted (MilliQwater) dye solutions— 15 ml (SY–Sunset Yellow; OII–Orange II; MO–Methyl Orange; MB–Methylene Blue; MV–Methyl Violet).	**DBD plane-to-plane plasma system (Gas–liquid)** Pulse Voltage: 23.8 to 31.4 kV. Pulse Frequency: 100 to 300 Hz. Treatment time: 2–20 min.	In NSP-DPD treatment >99.5% degradation was achieved in all the dye solutions within 20 min. Degradation kinetic rate: MV > MO > MB > OII > SY	(Meropoulis et al., 2021)
Methylene blue + deionized water (50 ml)	**Double chamber DBD plasma** V_{max}: 6 kV (18 W); Treatment time: up to 100 min. Gas: air and O_2; Flow rate: 60 std. cm^3	O_2 plasma–99.98% reduction after 20 min. Air-plasma–85.3% reduction after 100 min.	(Wang et al., 2017)
MB–170 mL with varying conductivity from 1 µS/cm to 1000 µS/cm.	**Plasma bubbling (wire plate type)** Frequency: 500 Hz Voltage: 30 kV Pulse width and rise time: 100 ns and 50 ns respectively. Gas: Dry air; Flow rate: 2 L/min.	95.7% discoloration within 10 min (average treatment power Pav=1.13 W). 93.6% discoloration within 20 min (Pav= 0.42 W).	(Abdelaziz et al., 2018)
MB solution–100 mL with the concentration of 20 mg/L.	NTP discharge with pyrite Gas: O_2, N_2, and Ar. Flow rate: 2 L/min Power: up to 70 W Treatment time: 60 min	At high discharge current (16.97 mA and 26.82 mA) N2 plasma degraded MB solution faster. While, at low current O2 plasma found to be effective. Also in acidic environment pyrite combined with O_2 plasma produced more OH species and improved MB degradation.	(Benetoli et al., 2012)

(Continued)

TABLE 8.1 (Continued)
Plasma Technology Applications in Wastewater Treatment

Sample	Treatment Condition	Experimental Results	Reference
MO—Methyl Orange with the conductivity of s 1700 μS/cm	**Multi-needle plasma reactor** Pulse voltage: 22 to 30 kV. Frequency: 50 Hz. Treatment time: 3 to 15 min. Gas: O_2, Air, N_2, and Ar.	Overall degradation efficiency with respect to the carrier gases was in the following order: O_2>Air>N_2>Ar **For optimum treatment** Needle to sample distance: 1 mm Sample to ground electrode distance: 28 mm; No. of electrodes (needles): 24	(B. Sun et al., 2012)
CV—Crystal violet (10 μM)	**RF atmospheric pressure plasma jet** Gas: Ar, Ar+1% O2, Ar+1% air and Ar+ 0.27% H_2O Air flow rate: 1.5 std. lit/min.	Degradation efficiency with respect to the carrier gases was in the following order: Ar > Ar+1% O_2 > Ar+1% air > Ar+0.27% H_2O	(Taghvaei et al., 2019)
Antibiotics (cefixime) in water (50 ml); Concentration: 50 mg/L, 100 mg/L and 200 mg/L	**Plasma bubbling** Power: 3.5–9.2 W Treatment time: up to 30 min Flow rate: 2.0 Std. Litre/min	**Antibiotics degradation after 30 min:** O_2 Plasma→~ 99.8% Air Plasma→ ~ 94.8% Ar plasma→~73.4% N_2 Plasma→~ 45.3%	(Zhang et al., 2021)
Textile effluent (Remazol blue RGB)— 50 ppm, 100 ppm and 150 ppm. Textile effluent (1:1 diluted with water)	**Microplasma reactor–capillary needles** Voltage: ~15 kV. Gas (air) flow rate: 1 std.litre/min.	99.4% degradation achieved in RB RGB solution (50 ppm) within 16 min. Mineralization—22.4%.	(Meiyazhagan et al., 2020)
25 ml water/meat effluent containing 10 mg/L antibiotics. (i) OFX—ofloxacin (ii) CFX—ciprofloxacin	**Atmospherics air DBD cold plasma—ACP** Input voltage: 70–80 kV. Treatment time: 5–25 min. Temperature: ~18 °C	CFX degradation in water: 75% and 89% at 70 kV and 80 kV. OFX degradation in water: 88% and 92% at 70 kV and 80 kV.	(Sarangapani et al., 2019)

Escherichia coli	**DBD-ACP** Plasma intensity: up to 1.77 kJ/cm² Treatment time: 4–14 min	Plasma at 0.71 kJ/cm² caused >3 log10 CFU/ml reduction in Escherichia coli.	(Liao et al., 2021)
Pseudomonas Aeruginosa Escherichia coli	**Microfluidic plasma reactor** Sample flow rate: 35 to 100 µL/min. Gas: Air (2 bars); Peak voltage:~10 kV	No survival after 5 s plasma treatment.	(Patinglag et al., 2021)
Acidithiobacillus ferrooxidans and Legionella gratiana	**40 pin electrodes plasma reactor** Gas: compressed air; Voltage: ~ 600 V; Treatment time: 0, 20, 40, and 60 s	A. ferrooxidans: 6 log within 40 s. L. gratiana 6 log reduction within 20 s.	(Johnson et al., 2016)
Amoxicillin	**DBD–falling liquid film (DBD-FLF)** Power: 2 W Treatment time: 10–30 min	100% degradation in 10 min.	(Monica Magureanu et al., 2015)
Ampicillin		100% degradation in 30 min.	
Oxacillin		100% degradation in 30 min.	
Sulfadiazine	**Wetted-wall Corona (WWC) and DBD-FLF** Power: 100 W Treatment time: 15–27 min.	99% degradation of sulfadiazine in corona discharge plasma after 27 min. 87% degradation of sulfadiazine in DBD-FLF plasma after 15 min.	(S. P. Rong et al., 2014; S. Rong & Sun, 2014)
Tetracycline	**Corona–gas bubbling** Power: 36 W; Treatment time: 24 min	61.9% degradation of tetracycline after 24 min treatment.	(He et al., 2014)
Carbamazepine	**DBD, DBD-FLF, and DBD-rotating drum reactor (DBD-RDR)** Power: 12 W and 500 W Treatment time: 30–60 min.	98%, 94%, and 90.7% degradation of carbamazepine in DBD (500 W), DBD-FLF (12 W), and DBD-RDR (500 W) plasma after 30 min, 60 min and 60 min respectively.	(Krause et al., 2009, 2011; Liu et al., 2012)
Clofibric acid	**DBD and DBD-RDR** Power: 500 W; Treatment time: 30 min.	Both plasma discharge reduced the clofibric acid by 100% within 30 min	(Krause et al., 2009, 2011)
Iopromide	**DBD and DBD-RDR** Power: 500 W; Treatment time: 30–60 min.	DBD and DBD-RDR plasma treatment of 30 min and 60 min removed 99% and 98% of iopromide respectively.	

(Continued)

TABLE 8.1 (Continued)
Plasma Technology Applications in Wastewater Treatment

Sample	Treatment Condition	Experimental Results	Reference
Pentoxifylline	**DBD-FLF** Power: 1.2 W; Treatment time: 60 min.	92% pentoxifylline was removed within 1hr of treatment	(Monica Magureanu et al., 2010)
Enalapril	**DBD-FLF** Power: 2 W; Treatment time: 120 min.	99.4% degradation	(Monica Magureanu et al., 2013)
Diclofenac	**Corona on water surface** Power: 24 W; Treatment time: 15 min.	100% degradation	(Dobrin et al., 2013)
Ibuprofen	**WWC discharge and Corona–liquid shower** Power: WWC–3 W; and corona liquid shower–250 W; Treatment time: 30 min and 80 min.	91.7% and 80% degradation achieved in WWC (80 min) and corona liquid shower (30 min) respectively.	(Zeng et al., 2015)
Paracetamol	**Corona–liquid shower (CLS)** Power: 250 W; Treatment time: 20 min.	100% degradation of paracetamol after 20 min treatment	(Panorel et al., 2013a)
Indomethacin & β-Oestradiol	**CLS plasma** Power: 120–250 W Treatment time: 5–30 min.	Indomethacin (90 mg/L): 100% degradation within 5 min treatment at 250 W. β-Oestradiol: 70% degradation within 20 min treatment at 120 W	(Panorel et al., 2013b)
Enrofloxacin (ENRO)	Nanosecond pulsed DBD plasma (gas–liquid discharge); Pulse voltage: ~26 kV; Frequency: 200 Hz; Gas: air, O_2, N_2; Flow rate: 1L/min; Treatment times: 2 to 20 min	ENRO was fully degraded at 26.2 kV and 23.4 kV voltage levels within 20 min of treatment time	(Aggelopoulos et al., 2020)
Diclofenac (conc. 50 mg/L) Conductivity: 250 µS/cm pH: 6	**Pulsed Corona on water surface** Max. voltage: ~20 kV; Gas: Oxygen; Flow rate: 1L/min	100% degradation of within 15 min.	(Dobrin et al., 2013)

is difficult to investigate and analyze the plasma characteristics and its components' influence for any particular experiment. Hence, researchers perform experiments in the trial and error method to obtain the desirable plasma characteristic or to get the required treatment outputs. Furthermore, predicting the plasma components becomes more difficult as it changes based on the operational parameters, such as supplied voltage, electrode distance, pulse repetition rate, gas composition, and air flow rate. Though the earlier research findings can be used as reference for future studies, quantitative assessment of plasma components only can provide reliable prediction of the processes.

Current advances in simulation techniques along with the technological assistance of the high-end computers (hardware, software) enable the complicated plasma process modeling possible to greater depth. At low cost and shorter time period, realistic simulations can be done with CFD techniques. These simulations provide a systematic assessment of the effects of changes in plasma operational and design parameters on the reactive species production. Further, with the suitable reaction kinetics and equations any treatment result can be predicted before the experiment has been done. The simulation models can also deliver detailed space-time data that are often difficult or impossible to obtain experimentally but are important for the understanding of physical process involved in plasma production (Ramshaw & Chang, 1992). The simulation of plasma at any given conditions through fluid or kinetic approach using CFD simulation software for forecasting its nature and interaction with targeted components is called 'computational plasma dynamics'.

In this computation, physical processes such as fluid dynamics, thermodynamics, and chemical reactions of plasma species will be considered. Among the three physical process of CFD, the fluid dynamics deal with the plasma flow properties with respect to mass, momentum, and energy conservation. While, the thermodynamics deal with the heat transfer between the plasma particles along with the work done and volume change caused by pressure (i.e., in the case of low pressure plasma). Finally, the reactive species generation and destruction can be determined by chemical reaction equations. All the physical processes involved in CFD are either evolving (at its own time-space scales) or in equilibrium (Yu et al., 1998).

The fluid approach, plasma is considered as a continuum where the governing equations, such as mass, momentum, and energy are solved using the Maxwellian equation for the velocity distribution function. In the kinetic approach, each element in plasma is considered a separate entity; the methods used mostly for solving the kinetic approaches are the Monte Carlo collision and particle-in-cell technique (Manoharan & Radhakrishnan, 2021). In either ways plasma modeling can be performed for predicting the plasma wastewater pollutant degradation at different operating conditions. There are many softwares available such as COMSOL Multiphysics, Autodesk CFD, SimScale, Ansys, OpenFOAM, Simcenter, Flowsquare, ParaView, Altair, SOLIDWORKS for CFD modeling. However, COMSOL Multiphysics provides a separate module along with multiphysics support opportunity to effectively model and simulate the plasma at different process conditions. Thus, section 8.4 mainly focuses only on the computational plasma dynamics approaches using COMSOL software as the reference. However, a major part of the discussion (i.e., equations, data collection) will be common for all CFD modeling platforms.

8.4 COMPUTATIONAL PLASMA DYNAMICS APPROACH

Before starting with CPD, it is essential to understand the basic concepts of CFD. In CFD, the physical aspects of fluid flow behavior are governed by the fundamental equations, such as mass, momentum, and energy conservation. Among these, the first law states that the mass can neither be created nor be destroyed; hence, mass entered inside the models will always be equal to the mass that went out of the system. This can be explained through the continuity equation [Eq. 8.1] for any CFD model:

$$\frac{\partial \rho}{\partial t} + \nabla.(\rho u) = 0 \qquad (8.1)$$

where ρ – density, t – time, and u – flow velocity. Whereas conservation of momentum is based on Newton's second law of motion [Eq. 8.2], which states that the total force acting on a body is equal to mass and acceleration of that body.

$$\rho \frac{dv}{dt} = -\nabla P + \rho f + F \qquad (8.2)$$

where the pressure force (P), particle force (f), and viscous forces act together and equalize the density and acceleration of the system. The conservation of energy is solved (continuity equation form) in CFD only when the density and the temperature of the fluid change during processing. In order to deal with both mass and momentum in a laminar flow interface, Navier–Stoke's equation is used in plasma modeling, which provides the fluid velocity along with the gas temperature of the system.

Plasma modeling is a multi-physics approach as it combines elements of reaction engineering, fluid mechanics, physical kinetics, electromagnetics, statistical physics, heat and mass transfer. The computed results are the solutions of multi-physics coupling between different physics interfaces of the built model. Software plasma modules contain simplified physical interfaces that are dedicated to modeling low-temperature plasma. These interfaces consist of all the mandatory tools to model plasma discharges by computing the electron transport properties and source coefficients from the provided cross section data using Boltzmann's equation, two-term approximation equations. With the available interface support, the data regarding discharge characteristics of plasma for any given input (operating) condition can be obtained (Ersion, 2013).

8.4.1 SELECTION OF SPACE DIMENSION

All plasma modeling starts with the selection of space dimension where the required geometrical of the plasma system can be drawn in 1D, 2D, and 3D (Fig. 8.2). The complexity and computation time of the modeling increase based on the increment in the dimension.

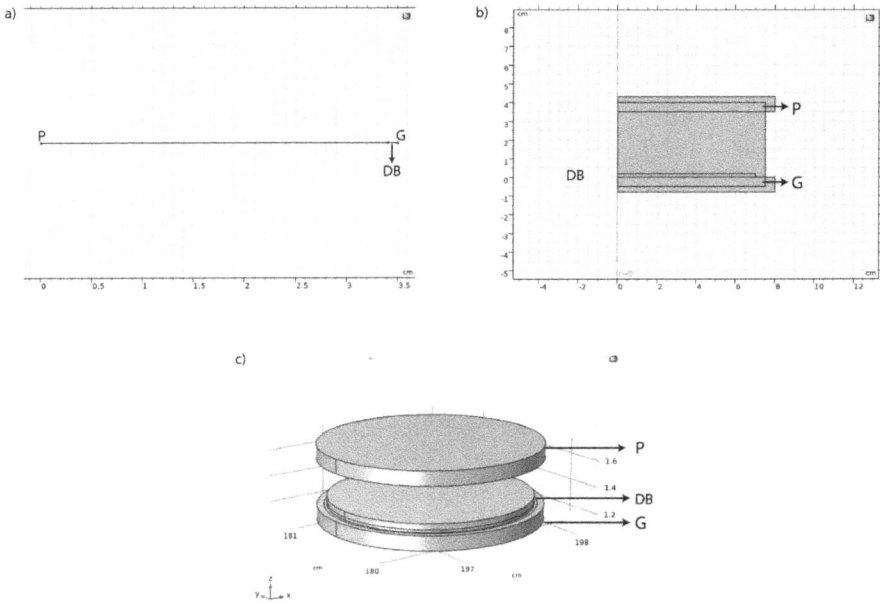

FIGURE 8.2 Different space dimension of a plasma system (a) 1D view, (b) 2D symmetry view, (c) 3D view (P – high voltage power electrode; G – ground electrode; DB – dielectric barrier).

8.4.2 Selection of Physics

Software build plasma physics contains various interfaces for modeling both thermal and non-thermal plasma. An example is plasma physics in COMSOL contains seven interfaces (plasma, inductively coupled plasma, microwave plasma, corona discharge, equilibrium discharge, species transport, and electron break down detection) each one of them has its unique in-build setup for modeling various plasma processes. Through multi-physics support, it is also possible to couple drift diffusion, heavy species transport, and charge transport to any of the selected plasma interfaces. However, some of the interfaces are already coupled with these existing interfaces (i.e., corona discharge interface).

8.4.2.1 Drift Diffusion

Using the drift diffusion interface, density and mean energy of the electrons can be computed for the given plasma conditions. The electron transport properties required for the drift diffusion is computed using Boltzmann's equation along with the two-term approximation interface. Through proper boundary conditions, it is possible to take account of the secondary and thermionic emission along with the wall losses into the model. The equations used for computing electron density (n_e – Eq. 8.3), electron energy density (n_ε – Eq. 8.4) and mean energy ($\bar\varepsilon$ – Eq. 8.5) are as follows:

$$\frac{\partial}{\partial t}\left(n_e\right)+\nabla\left[-\left(\mu_e \cdot E\right)n_e-\nabla\left(D_e n_e\right)\right]=R_e-\left(u\cdot\nabla\right)n_e \tag{8.3}$$

$$\frac{\partial}{\partial t}\left(n_\varepsilon\right)+\nabla\left[-\left(\mu_\varepsilon \cdot E\right)n_\varepsilon-\nabla\left(D_\varepsilon n_\varepsilon\right)\right]+\mathrm{E}\cdot\left[-\left(\mu_e\cdot E\right)n_e-\nabla\left(D_e n_e\right)\right]$$
$$=S_{\mathrm{en}}-\left(u\cdot\nabla\right)n_\varepsilon+\left(Q+Q_{\mathrm{gen}}\right)/q \tag{8.4}$$

$$\bar{\varepsilon}=\frac{n_\varepsilon}{n_e} \tag{8.5}$$

where μ_e – electron mobility; D_e – electron diffusivity; E – electric field; R_e – electron rate expression; u – neutral fluid velocity (negligible for electrons); μ_ε – electron energy mobility; D_ε – electron energy diffusivity; S_{en} – change of energy due to inelastic collisions; Q – external heat source; Q_{gen} – generalized heat source.

8.4.2.2 Heavy Species Transport

This interface helps in computing mass fraction of heavy species present in the plasma. Though different methods are available for modeling the mass transport, Maxwell–Stefan is preferred due to its total mass conservation and auxiliary constraints satisfying ability. However, more than six species cannot be computed using Maxwell–Stefan equation due to computational difficulties. In addition, the equation itself demands more computational resources and also assumes one particular temperature for all heavy species. Thus, the interface provides mixture averaged and Fick's law diffusion models for the species transport. Though Fick's law is a simple model to compute, mixture averaged is preferred due to its accuracy. When a reaction of reactive species (i.e., $e+O_2 \rightarrow e+O+O$) is considered for plasma modeling that reaction can be specified using cross-sectional data, Arrhenius parameters, rate constant, and lookup data. When cross-sectional data are used to get the source coefficient, electron energy distribution function (EEDF) needs to be selected in the modeling (Maxwellian [Eq. 8.6], Druyvesteyn [Eq. 8.7], and generalized [Eq. 8.8] EEDF are the available functions).

$$f\left(\varepsilon\right)=\varphi^{-3/2}\beta_1\exp\left(-\left(\frac{\varepsilon\beta_2}{\varphi}\right)\right) \tag{8.6}$$

where
φ – mean electron energy (eV), ε – electron energy (eV);

$$\beta_1=\Gamma(5/2)^{3/2}\,\Gamma(3/2)^{-5/2};\beta_2=\Gamma(5/2)^{3/2}\,\Gamma(3/2)^{-1}$$

$$\Gamma(s)=\int_0^\infty u^{s-1}e^{-u}du$$

$$f(\varepsilon) = 2\varphi^{-3/2}\beta_1 \exp\left(-\left(\frac{\varepsilon\beta_2}{\varphi}\right)^2\right) \tag{8.7}$$

where

$$\beta_1 = \Gamma(5/4)^{3/2}\,\Gamma(3/4)^{-5/2}; \beta_2 = \Gamma(5/4)\Gamma(3/4)^{-1}$$

$$f(\varepsilon) = g\varphi^{-3/2}\beta_1 \exp\left(-\left(\frac{\varepsilon\beta_2}{\varphi}\right)^g\right) \tag{8.8}$$

where

$$\beta_1 = \Gamma(5/2g)^{3/2}\,\Gamma(3/2g)^{-5/2}; \beta_2 = \Gamma(5/2g)\Gamma(3/2g)^{-1}; 1 \le g \ge 2$$

For all the above mentioned cases, the rate constant (k_f) will be computed using [Eq. 8.9].

$$k_f = \gamma\int_0^{\infty}\varepsilon\sigma_k(\varepsilon)f(\varepsilon)d\varepsilon \tag{8.9}$$

where
$\gamma = (2q/m_e)^{0.5}$, $(C^{0.5}/kg^{0.5})$; m_e – electron mass (kg); ε – electron energy (V); σ_k – collision cross section (m²); f – EEDF.

8.4.2.3 Charge Transport

This interface calculates the density of charged particles in a background gas assuming that migration is the dominant mode of transport. The charge transport interface best suits corona discharge plasma when coupled with the electrostatic interface; using the interface, both electric field and potentials can be computed. The charge density equation can be obtained by combining current density, charge conservation, and Poisson's equations [Eq. 8.10] as given below.

$$\mu_i\left(\frac{\rho^2}{\varepsilon_0} - \nabla V \cdot \nabla\rho\right) + \rho\nabla V \cdot \nabla\mu_i + \nabla\rho \cdot u = S \tag{8.10}$$

where J – current density; ρ – space charge density; S – current source; Z_q – charge number; E – electric field; V – electric field; μ_i – ion mobility; u – neutral fluid velocity vector (Navier–Stokes interface)

8.4.3   Study Selection

The plasma simulation can be time-dependent, stationary, and pre-set study. Time-dependent studies can be carried out when plasma field variables are changing over the time (i.e., electron density change over the time). When there is no change in the plasma varies with respect to time, stationary studies are preferable. While, the software pre-set studies are suited for limited data extraction as these programs allow the user to execute only the EEDF without expensive computational procedures. Apart from that, empty studies can be used for getting customized computed plasma data.

8.4.4   Defining the Parameters and Variables

Based on the plasma type and operational conditions, the parameters need to be defined. Power level, frequency, pressure, and temperature are some of the basic parameters that are used in plasma modeling. Any alpha-numerical can be used to denote a parameter with the standard unit expression in the adjutant column (Fig. 8.3-a) and the values given to each parameter can be modified later according to treatment conditions. Treatment variable can either be defined as global variable or as local variable.

Global variables are independent variable expressions that is not affected by component, geometric domain, boundary (edges or point) of a model, while the local variables are influenced by these factors. When defining the variable, it is preferable

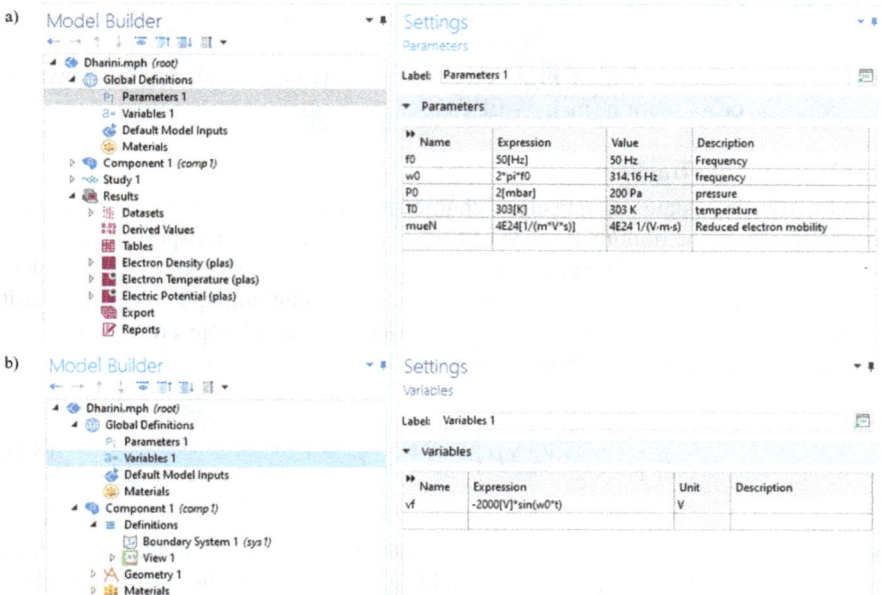

FIGURE 8.3   COMSOL model builder (a) settings for parameter (b) settings for variable (source: COMSOL Multiphysics 5.6).

to define them under components of the model as it minimizes the variable name errors. Log of electron density (N_e), log of electron energy density (E_n) and electrical potential (V) are some of the dependent variables of plasma modeling, which are built-in with the plasma physics interface.

8.4.5 GEOMETRY BUILDING

A plasma system can be modeled by building its 1D, 2D, or 3D geometry based on the selected space dimension. Building a 1D model involves coordinates or interval-based approaches for drawing the lines and points. Although, 2D and 3D geometries can be drawn by fixing the dimensions and coordinates of individual parts with the support of geometry toolbar (Fig 8.4-a). Apart from building the geometry in the COMSOL software, it is also possible to import the geometries from external software such as AutoCAD, SOLID works (Geometry from different COMSOL files can also be imported). The source files that can be read in the COMSOL importer mesh file, COMSOL Multiphysics file, DXF/CAD file, ECAD file.

8.4.6 MATERIAL SELECTION

Based on the wastewater treatment system (plasma unit) construction, different martials can be selected from 'Material toolbar' (Fig. 8.4-b) to build a model. The selected materials can be assigned to different geometrical parts (domains) of the

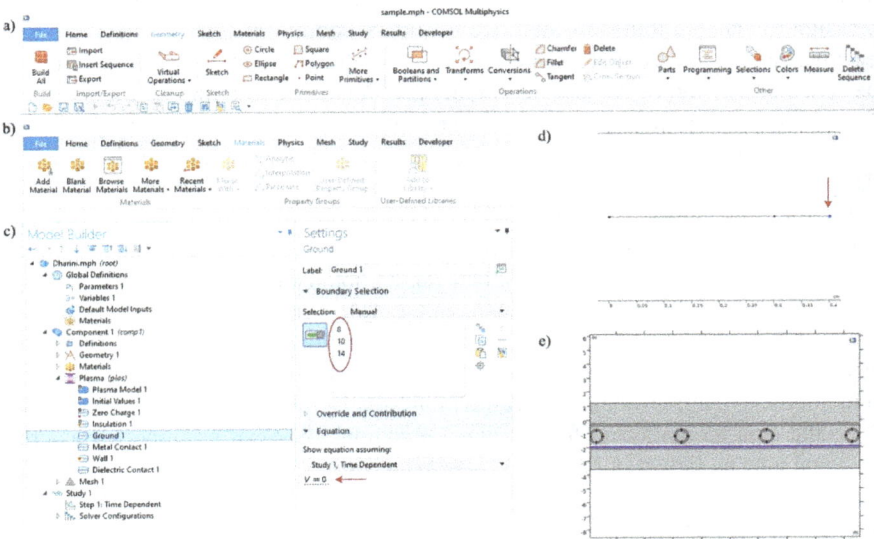

FIGURE 8.4 COMSOL Multiphysics workspace. (a) Geometry toolbar. (b) Material selection toolbar. (c) Plasma physics-boundaries selection. (d) ID geometry graphical view. (e) 2D geometry graphical view.

plasma model by manual domain selection. For each selected material, a set of in-build properties is available (i.e., equilibrium discharge, electromagnetic model) in the material properties' node. In addition, the material type (solid, liquid) also can be defined under material properties' node, which determines the behavior of the material and also helps in interpretation of material properties when mesh (section 8.4.8) undergoes deformation. Furthermore, the material content also provides a list of material properties that are either predefined for selected material or essential for selected plasma physics to run the model. The values and unit of the material properties can be edited with proper unit substitutions to define or modify a particular material nature. Defining properties and providing physics (plasma) demanding material data could avoid unnecessary errors in the computation process.

8.4.7 BOUNDARY CONDITIONS UNDER SELECTED PHYSICS

Brief details on the model physics and their governing equations are given in section 8.4.2. Once the physics is selected, the next step is to select the boundary condition inside the model builder interface. The boundary conditions vary according to the selected physics (Table 8.2) and once the boundary conditions are satisfied for the given model, it can be used for computing data. For example, drift diffusion interface contains equations for computing electron transport using the input data such as electric potential, collision power loss. Although, the heavy species transport interface computes the mass fraction of heavy species present in the plasma. In the charge transport interface, drift and convective transport are defined by parameters such as model inputs (temperature, absolute pressure), electric potential, reduced ion mobility, charge number and velocity to solve ions density. However, it is necessary to assign the charge transport interface boundary conditions for the selected plasma reactor geometry. Nevertheless, the boundary condition selection varies according to the space dimension of the selected model geometry. For example, in Fig. 8.4-c, the 'Ground 1' is selected to represent the ground electrode of the plasma system; this 'Ground 1' for an 1D model is shown in Fig. 8.4-d (red arrow) as a point, while for 2D symmetrical model (Fig. 8.4-e), it is shown as boundaries (8, 10, 14).

In order to simulate a plasma reaction, the data regarding various reactions (electron impact reactions, general reactions, and surface reactions) and reactive species needs to be known. These reactions data can be obtained through the LXCat (www.lxcat.net) website. In wastewater treatment the water and plasma interaction happens in four different forms as seen in Fig. 8.1. Based on the zone of contact the reactive species generated inside the water treatment system also varies. Hence, it is essential to include the reactions that are specific to plasma and water interactions (Table 8.3). Once the data are collected for the given reactions, they can be imported as cross-sectional data as shown in Fig. 8.5.

8.4.8 MESHING

The process of enabling the discretion of a given geometry into small elements of defined shapes is called meshing (Fig. 8.6). The mesh can be of user-controlled or

TABLE 8.2
Boundary Conditions at Different Plasma Model Interfaces

Drift Diffusion	Heavy Species Transport	Charge Transport
Initial value: $n_{e,0}$–Initial electron density (1/m³). e_0–Initial mean electron energy (SI unit: V).	Flux: $-n \bullet \rho w_k \left(V_k + u \right) = \Gamma_k$ where, ρ – density (kg/m³); w_k – mass fraction of kth reactive species. u – fluid velocity (m/s) V_k – diffusive velocity of kth reactive species (m/s);	Initial value: feed space charge density (C/m³) value.
Wall: both electron loss and gain occurs in this boundary. Electron gain is due to secondary and thermionic emissions; while electron loss occurs due to net flux and random motions.	Inlet: input value can be of any of the followings→ mass fraction, mole fraction, molar concentration or number density	No flux: $-n \cdot J = 0$; where J-current density, 'nJ' represent the normal component current density. (Note: J value of the charge carriers is zero in this boundary)
Insulation: $-n \bullet \Gamma_e = 0$ $-n \bullet \Gamma_\varepsilon = 0$ 'n' present normal component of Γ_e and Γ_ε	Outflow: $-n.j_K = 0$ j_k – diffusion flux of kth reactive species (Assumption: convective mass transfer in boundary)	Source: current source (A/m³) or reaction rate (mol/m³.s)
Electron density (n_e) and energy (e) boundary: Fixing of n_e and e to a certain value. (Note: this specific boundary condition is avoided to facilitate wall boundary condition)		Space-charge density: $\rho = \rho_0$ (or) $\rho = n_0 e$; where, n_0 – number density (1/m³), ρ_0 – space-charge density (C/m³)

physics-controlled mesh; these mesh (element) size can be controlled in both the methods to solve the physics. In ID model meshing, the geometry intervals (domains) are further divided into smaller intervals, while the boundaries of the model are represented by vertex elements. Elements of specific shapes will be used for modeling 2D and 3D geometry. In 2D meshing, triangular or quadrilateral shaped meshing elements will be used to discrete the geometry while the boundaries are divided by mesh element sides (edges). Similarly, for 3D geometry, mesh element shapes such as tetrahedral, prism, pyramid, or hexahedral are available.

TABLE 8.3

Reactive Species Interaction with Water (Wastewater) under Different Plasma Discharge Conditions

Plasma Discharge above the Water Surface	Direct Discharge in Aqueous Phase	Plasma Bubbling inside Aqueous Phase	Discharge in Hybrid Gas-Liquid System
$e^- + O_2 \rightarrow O\cdot + O\cdot + e^-$	$H_2O \rightarrow H\cdot + HO\cdot$	$O_2 + O_2 \rightarrow O_3 + O\cdot$	$2HO\cdot \rightarrow H_2O_2$
$e^- + H_2O \rightarrow H\cdot + HO\cdot + e^-$	$H_2O \rightarrow \frac{1}{2}(H_2O_2) + \frac{1}{2}(H_2)$	$O_3 \rightarrow O_2 + O\cdot + e$	$2HO\cdot_{(g)} + M \rightarrow H_2O_{2(g)} + M$
$e^- + O_2 \rightarrow O_2^+ + 2e^-$	$2H_2O \rightarrow H_3O^+ + HO\cdot + e^-$	$H_2O + O\cdot \rightarrow 2OH\cdot$	$2HO\cdot_{(aq)} \rightarrow H_2O_{2(aq)}$
$e^- + O_2 + M \rightarrow O_2^- + M$	$H\cdot + O_2 \rightarrow HO_2\cdot$	$O_3 + NO \rightarrow NO_2 + O_2$	$2HO\cdot_{(int)} \rightarrow H_2O_{2(int)}$
$O\cdot + O_2 + M \rightarrow O_3 + M$	$H + H_2O_2 \rightarrow H_2O + HO\cdot$	$O_3 + NO \rightarrow NO_2 + O_2$	aq – aqueous; g – gas; int – air–liquid interface.
$e^- + 2H_2O \rightarrow H_2O_2 + H_2 + e^-$	$HO + H_2O_2 \rightarrow H_2O + HO_2\cdot$	$O_2 + N \rightarrow NO + N$	Note: the interaction of reactive species with water under each category is not restricted to the given reactions.
$HO\cdot + H_2O_2 \rightarrow H_2O + HO_2\cdot$	$e^-_{eq} + HO\cdot \rightarrow OH^-$	$2NO + O_2 \rightarrow 2NO_2$	
$O + H_2O \rightarrow HO\cdot + HO\cdot$	$e^-_{eq} + H + H_2O \rightarrow OH^- + H_2$	$NO_2 + OH \rightarrow HNO_3$	
$O_3 + H_2O_2 \rightarrow HO\cdot + O_2 + HO_2\cdot$	$e^-_{eq} + H_2O_2 \rightarrow HO\cdot + OH^-$	$O_3 + H_2O \rightarrow H_2O_2 + O_2$	
$O_3 + HO_2\cdot \rightarrow HO\cdot + O_2 + O_2^-$	$H\cdot + HO\cdot \rightarrow H_2O$	$NO + OH\cdot \rightarrow HNO_2$	
$O_3 + h \rightarrow H_2O + H_2O_2 + O_2$	$2HO\cdot \rightarrow H_2O_2$	$NO_2 + NO_3 \rightarrow N_2O_5$	
$H_2O_2 + h \rightarrow 2HO\cdot$	$2HO_2\cdot \rightarrow H_2O_2 + O_2$	$3NO_2^- + 3H^+ \rightarrow 2NO + NO_3^- + H_3O^+$	
$e^- + N_2 \rightarrow 2N\cdot + e^-$	$H\cdot + HO_2\cdot \rightarrow H_2O_2$	$HNO_2 + OH\cdot \rightarrow NO_2 + H_2O$	
$N\cdot + O\cdot \rightarrow NO$	$2H\cdot \rightarrow H_2$		
$O\cdot + NO \rightarrow NO_2$	$HO_2\cdot + HO\cdot \rightarrow H_2O_2 + O_2$		
	$H_3O^+ + OH^- \rightarrow 2H_2O$		

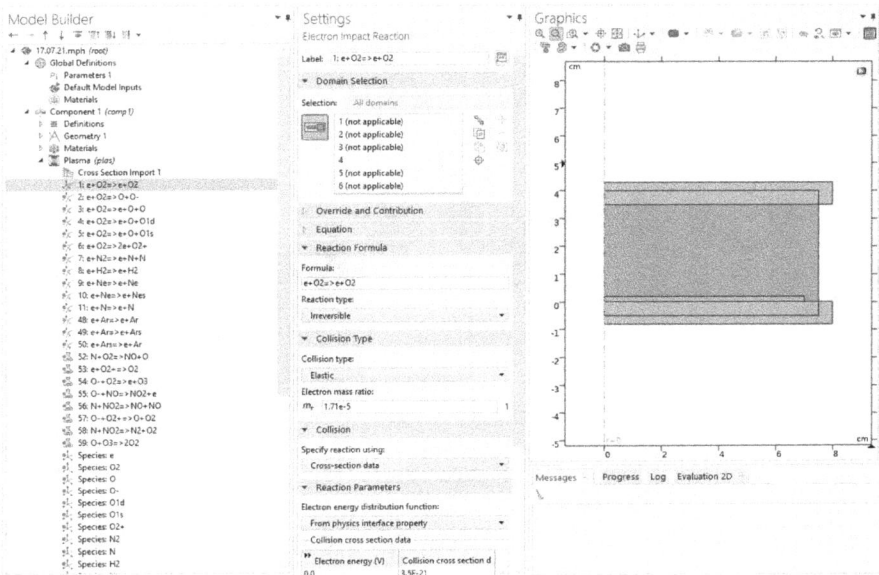

FIGURE 8.5   Cross-section data import for reactive species (source: COMSOL Multiphysics).

8.4.9   STUDY AND SOLUTION COMPUTING

The study node available in the software (i.e., COMSOL multiphysics) provides the braches of sub-nodes to explore characteristics of plasma at different operating conditions (Fig. 8.7). The study and the study steps available in the interface enables the control on the equation, physics and mesh used in the model. Based on the study that was selected during the initial stage of the modeling process decides the final study interface. As an example, if stationary study is selected for a model, then the study node will only provide the option to compute the result for stationary study steps. However, the interface consists of certain general studies that are applicable for all study types. While the pre-set studies compute all applicable studies automatically for the selected physics or multi-physics (when coupled) of the model. However, for professional with high computational modeling skills, the empty studies available in the modeling interface enables new study approaches.

8.4.10   COMPUTATIONAL PLASMA DYNAMICS FOR WASTEWATER TREATMENT AND RESULT INTERPRETATION

Sun et al's (2016) study on reactive blue 19 (RB-19) degradation showed 59.9%, 49.6%, and 89.8% dye degradation when using Ar, air, and O_2 as carrier gases at 100 kJ/L energy density. Though different reactive species plays their role in dye degradation, ozone produced during plasma treatment has a major impact. Hence, a 1D model was developed using COMSOL® for determining the O_3 contribution on RB-19 oxidation. Since, the discharge was above the water surface, the equations (Eq. 8.11)

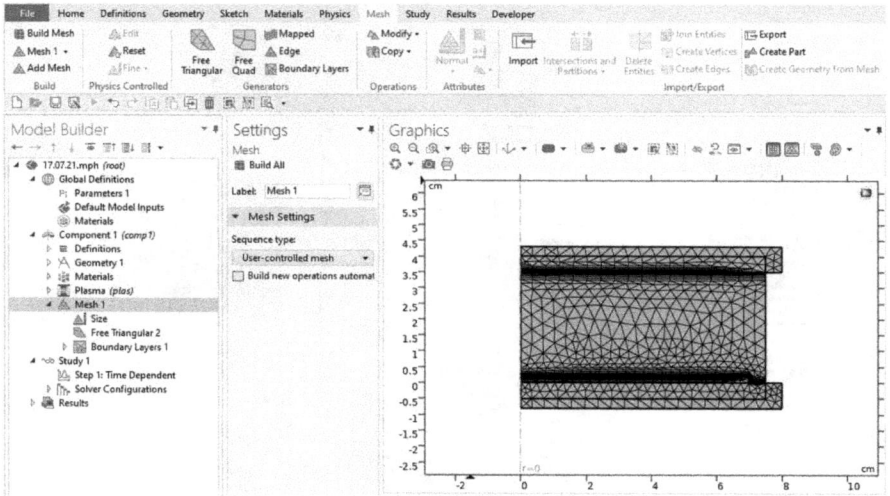

FIGURE 8.6 Meshing operation (source: COMSOL Multiphysics).

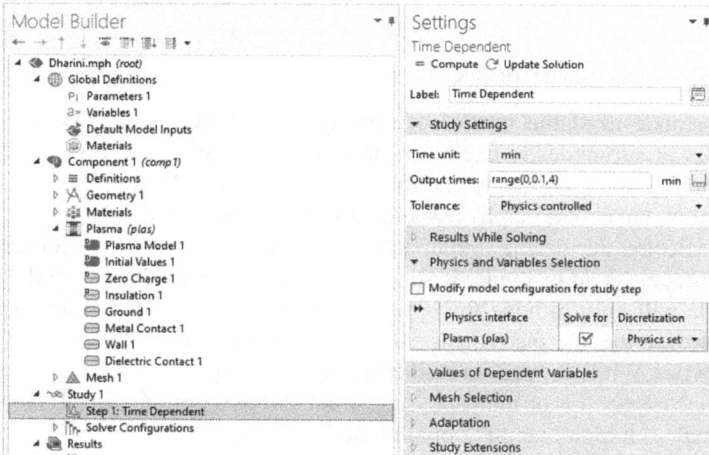

FIGURE 8.7 Study interface for time-dependent plasma model (source: COMSOL Multiphysics).

for convection-diffusion were solved to predict the dye and ozone concentrations in the liquid phase.

$$\nabla \cdot \left(-D_c \nabla C\right) = R_{cdye} - u \cdot \nabla C \tag{8.11}$$

The diffusion coefficient D_c was calculated by making various assumptions (through the heptanol dispersion co-efficient model). However, the D_c values for which the simulation results were close to the experimental result were presented in the study. In Eq. 8.1, the term R_{cdye} indicates the consumption of O_3 (mol m^{-3} s^{-1}) during the dye degradation process that can be obtained from the Eq. 8.12 (k-reaction constant, C_{O3}-ozone concentration in liquid, and C_{dye}-dye concentration in liquid). In addition, Fick's law was used to solve the boundary condition of the system Eq. 8.13.

$$R_{cdye} = -k \cdot C_{O_3} C_{dye} \qquad (8.12)$$

$$\frac{\partial C}{\partial t} = \nabla \left(D \nabla C \right) \qquad (8.13)$$

Compared to experimental results, simulation showed less degradation for all the treatment periods (5–15 min) due to the omission of non-ozone reactive species in the model. From the simulation model, the ozone contribution in the dye degradation was found to be 0.67 during the initial 5 min treatment and it was increased to 0.82 in the next 10 min of the treatment. Apart from predicting the end results, CFD can also be used for optimizing the operation parameters like voltage/potential difference between electrodes. EL-Tayeb et al. (2015) used a pin-to-plate corona discharge system to treat acid blue-25 dye contaminated wastewater with the focus of optimizing the electrodes' distance for better dye degradation. In this regard, a 3D simulation model was made to identify the corona onset voltage of the system using a software utilizing the integral equations technique (IET) to solve the partial differential equations. Using IET 3D electrostatic design electric field and potential were analyzed for the system at 15 kV input voltage. From the results, a 5 mm air gap was found to be effective for acid blue-25 degradation and caused 80% of dye within 10 min of treatment. The shaded plot obtained from simulation showed higher electric field distribution near the sharper end of the electrode pins than the other parts. Hence, a sharper electrode region was decided to be the high corona region (other details are not provided). However, due to the expensive computational procedures, it is not always possible to compute a 3D model. Therefore, adapting a suitable 2D symmetrical geometry can reduce the computational difficulties and simulation time to a greater extent. Ognier et al. (2009) used a 2D symmetrical geometry model (COMSOL) for computing the gas–liquid discharge process and tried to understand pollutant (1-heptanold, phenol) degradation mechanism in wastewater. For calculating the thickness (Eq. 8.14) of liquid film that comes in contact with plasma, the wastewater was assumed to be in laminar flow with parabolic velocity profile.

$$\bar{\delta} = \left(3 \frac{\mu}{\rho} \times \frac{R_e}{g} \right)^{1/3} \qquad (8.14)$$

where

$$R_e = \frac{\dot{m}}{\pi D \mu}$$, μ – viscosity, ρ – density, g – gravity, $\dot{m}$ – mass flow (kg/s), D – liquid

tube diameter.

From Eq. 8.14, the film thickness was found to be 0.14 mm (flow rate – 6 mL/min), while the gas velocity was computed using the Navier–Stokes equation. The pollutant and O_3 concentrations at liquid and gas phase were computed using the convection-diffusion equation at both aqueous and gas phase.

$$\nabla.(-D\nabla c) = R - u.\nabla c \tag{8.15}$$

The reactive species with shorter life span was considered in the gas phase and using Eq. 8.16 the oxidation reaction was computed.

$$R = -K_{app} \times C_{gas}^P \tag{8.16}$$

where C_{gas}^P – pollutant concentration in gas; K_{app} – reaction rate, which depends on the selected reactive species.

However, only O_3 species was considered in liquid phase and its concentration was calculated as the function of discharge energy density. The gas and liquid concentrations were assumed to be thermodynamically equilibrium at the boundary (Eq. 8.17).

$$C_{liq} = K_h C_{gas} \tag{8.17}$$

The computed result showed only 4% removal of 1-heptanol while the actual treatment removed up to 90% of the pollutant. Thus, it was found that the RS present in the gas phase alone cannot be considered for modeling as a major part of the pollutant damage was caused by the RS present (mixed) inside the liquid. However, when the assumptions (mass transfer in gas-liquid interface) were made along with suitable diffusion coefficient, the computation became effective and showed better results. The computed and calculated results for 1-heptanol, ethanol, phenol and acetic acid are 91%, 51%, 13%, and 7.5% and 88%, 40%, 20%, and 7%, respectively. It was also found that the O_3 present in the liquid caused 30% pollutant removal.

8.5 CONCLUSION

Though non-thermal plasma is gaining popularity in wastewater treatment, it is not commercialized due to its construction complexity and optimization expenses involved in large-scale operations. However, development in CPD techniques can solve research gaps between the lab and large-scale studies. Hence, it is possible to optimize and build plasma treatment plants for all types of wastewater treatments ranging from microbial removal to dye degradation. Further, CPD modeling itself can be improved by high-end software and hardware supports, which can reduce the computation time and enhance multi-physics coupling. Nevertheless, the computation accuracy is based on the input plasma species reaction that we are feeding through cross-sectional data. Thus, understanding the plasma reactive species generation and their mechanisms involved in wastewater treatment will improve the computation efficiency of the developed model.

REFERENCES

Abdelaziz, A. A., Ishijima, T., & Tizaoui, C. (2018). Development and characterization of a wire-plate air bubbling plasma for wastewater treatment using nanosecond pulsed high voltage. *Journal of Applied Physics*, *124*(5). https://doi.org/10.1063/1.5037107

Aggelopoulos, C. A., Meropoulis, S., Hatzisymeon, M., Lada, Z. G., & Rassias, G. (2020). Degradation of antibiotic enrofloxacin in water by gas-liquid nsp-DBD plasma: Parametric analysis, effect of H_2O_2 and CaO_2 additives and exploration of degradation mechanisms. *Chemical Engineering Journal*, *398*(February), 125622. https://doi.org/10.1016/j.cej.2020.125622

Anandharamakrishnan, C. (2013). Computational fluid dynamics applications in food processing. In C. Anandharamakrishnan (Ed.), *Computational Fluid Dynamics Applications in Food Processing*. Springer New York. https://doi.org/10.1007/978-1-4614-7990-1

Benetoli, L. O. de B., Cadorin, B. M., Baldissarelli, V. Z., Geremias, R., de Souza, I. G., & Debacher, N. A. (2012). Pyrite-enhanced methylene blue degradation in non-thermal plasma water treatment reactor. *Journal of Hazardous Materials*, *237–238*, 55–62. https://doi.org/10.1016/j.jhazmat.2012.07.067

Crini, G., & Lichtfouse, E. (2019). Advantages and disadvantages of techniques used for wastewater treatment. *Environmental Chemistry Letters*, *17*(1), 145–155. https://doi.org/10.1007/s10311-018-0785-9

Dobrin, D., Bradu, C., Magureanu, M., Mandache, N. B., & Parvulescu, V. I. (2013). Degradation of diclofenac in water using a pulsed corona discharge. *Chemical Engineering Journal*, *234*, 389–396. https://doi.org/10.1016/j.cej.2013.08.114

EL-Tayeb, A., El-Shazly, A. H., Elkady, M. F., & Abdel-Rahman, A. (2015). Decolorization of Acid blue 25 dye by non-thermal plasma advanced oxidation process for industrial wastewater treatment. *2015 IEEE 15th International Conference on Environment and Electrical Engineering, EEEIC 2015–Conference Proceedings*, 807–812. https://doi.org/10.1109/EEEIC.2015.7165268

Ersion, V. (2013). Introduction to Plasma Module. In *Plasma Module User's Guide*. https://doc.comsol.com/5.6/doc/com.comsol.help.plasma/PlasmaModuleUsersGuide.pdf

Ezugbe, E. O., & Rathilal, S. (2020). Membrane technologies in wastewater treatment: A review. *Membranes*, *10*(5), 89. https://doi.org/10.3390/membranes10050089

He, D., Sun, Y., Xin, L., & Feng, J. (2014). Aqueous tetracycline degradation by non-thermal plasma combined with nano-TiO_2. *Chemical Engineering Journal*, *258*, 18–25. https://doi.org/10.1016/j.cej.2014.07.089

Jiang, B., Zheng, J., Qiu, S., Wu, M., Zhang, Q., Yan, Z., & Xue, Q. (2014). Review on electrical discharge plasma technology for wastewater remediation. *Chemical Engineering Journal*, *236*, 348–368. https://doi.org/10.1016/j.cej.2013.09.090

Johnson, D. C., Bzdek, J. P., Fahrenbruck, C. R., Chandler, J. C., Bisha, B., Goodridge, L. D., & Hybertson, B. M. (2016). An innovative non-thermal plasma reactor to eliminate microorganisms in water. *Desalination and Water Treatment, 57*(18), 8097–8108. https://doi.org/10.1080/19443994.2015.1024752

Krause, H., Schweiger, B., Prinz, E., Kim, J., & Steinfeld, U. (2011). Degradation of persistent pharmaceuticals in aqueous solutions by a positive dielectric barrier discharge treatment. *Journal of Electrostatics, 69*(4), 333–338. https://doi.org/10.1016/j.elstat.2011.04.011

Krause, H., Schweiger, B., Schuhmacher, J., Scholl, S., & Steinfeld, U. (2009). Degradation of the endocrine disrupting chemicals (EDCs) carbamazepine, clofibric acid, and iopromide by corona discharge over water. *Chemosphere, 75*(2), 163–168. https://doi.org/10.1016/j.chemosphere.2008.12.020

Liao, X., Liu, D., Chen, S., Ye, X., & Ding, T. (2021). Degradation of antibiotic resistance contaminants in wastewater by atmospheric cold plasma: kinetics and mechanisms. *Environmental Technology (United Kingdom), 42*(1), 58–71. https://doi.org/10.1080/09593330.2019.1620866

Liu, Y., Mei, S., Iya-Sou, D., Cavadias, S., & Ognier, S. (2012). Carbamazepine removal from water by dielectric barrier discharge: Comparison of ex situ and in situ discharge on water. *Chemical Engineering and Processing: Process Intensification, 56*, 10–18. https://doi.org/10.1016/j.cep.2012.03.003

Magureanu, M., Bilea, F., Bradu, C., & Hong, D. (2021). A review on non-thermal plasma treatment of water contaminated with antibiotics. *Journal of Hazardous Materials, 417*, 125481. https://doi.org/10.1016/j.jhazmat.2021.125481

Magureanu, Monica, Dobrin, D., Mandache, N. B., Bradu, C., Medvedovici, A., & Parvulescu, V. I. (2013). The mechanism of plasma destruction of enalapril and related metabolites in water. *Plasma Processes and Polymers, 10*(5), 459–468. https://doi.org/10.1002/ppap.201200146

Magureanu, Monica, Mandache, N. B., & Parvulescu, V. I. (2015). Degradation of pharmaceutical compounds in water by non-thermal plasma treatment. *Water Research, 81*, 124–136. https://doi.org/10.1016/j.watres.2015.05.037

Magureanu, Monica, Piroi, D., Mandache, N. B., David, V., Medvedovici, A., & Parvulescu, V. I. (2010). Degradation of pharmaceutical compound pentoxifylline in water by non-thermal plasma treatment. *Water Research, 44*(11), 3445–3453. https://doi.org/10.1016/j.watres.2010.03.020

Mahendran, R., & Alagusundaram, & K. (2015). Uniform discharge characteristics of non-thermal plasma for superficial decontamination of bread slices. *International Journal of Agricultural Science and Research (IJASR), 5*(2), 209–212.

Manoharan, D., & Radhakrishnan, M. (2021). Computational cold plasma dynamics and its potential application in food processing. *Reviews in Chemical Engineering*. https://doi.org/10.1515/revce-2021-0005

Meiyazhagan, S., Yugeswaran, S., Ananthapadmanabhan, P. V., & Suresh, K. (2020). Process and kinetics of dye degradation using microplasma and its feasibility in textile effluent detoxification. *Journal of Water Process Engineering, 37*(July), 101519. https://doi.org/10.1016/j.jwpe.2020.101519

Meropoulis, S., Rassias, G., Bekiari, V., & Aggelopoulos, C. A. (2021). Structure-Degradation efficiency studies in the remediation of aqueous solutions of dyes using nanosecond-pulsed DBD plasma. *Separation and Purification Technology, 274*(March), 119031. https://doi.org/10.1016/j.seppur.2021.119031

Miklos, D. B., Remy, C., Jekel, M., Linden, K. G., Drewes, J. E., & Hübner, U. (2018). Evaluation of advanced oxidation processes for water and wastewater treatment–A critical review. *Water Research, 139*, 118–131. https://doi.org/10.1016/j.watres.2018.03.042

Ognier, S., Iya-Sou, D., Fourmond, C., & Cavadias, S. (2009). Analysis of mechanisms at the plasma-liquid interface in a gas-liquid discharge reactor used for treatment of polluted water. *Plasma Chemistry and Plasma Processing*, *29*(4), 261–273. https://doi.org/10.1007/s11090-009-9179-x

Pankaj, S. K., Wan, Z., & Keener, K. M. (2018). Effects of cold plasma on food quality: A review. *Foods*, *7*(1), 4. https://doi.org/10.3390/foods7010004

Panorel, I., Preis, S., Kornev, I., Hatakka, H., & Louhi-Kultanen, M. (2013a). Oxidation of aqueous paracetamol by pulsed corona discharge. *Ozone: Science and Engineering*, *35*(2), 116–124. https://doi.org/10.1080/01919512.2013.760415

Panorel, I., Preis, S., Kornev, I., Hatakka, H., & Louhi-Kultanen, M. (2013b). Oxidation of aqueous pharmaceuticals by pulsed corona discharge. *Environmental Technology (United Kingdom)*, *34*(7), 923–930. https://doi.org/10.1080/09593330.2012.722691

Patinglag, L., Melling, L. M., Whitehead, K. A., Sawtell, D., Iles, A., & Shaw, K. J. (2021). Non-thermal plasma-based inactivation of bacteria in water using a microfluidic reactor. *Water Research*, *201*(May), 117321. https://doi.org/10.1016/j.watres.2021.117321

Potluri, S., Sangeetha, K., Santhosh, R., Nivas, G., & Mahendran, R. (2018). Effect of low-pressure plasma on bamboo rice and its flour. *Journal of Food Processing and Preservation*, *42*(12), e13846. https://doi.org/10.1111/jfpp.13846

Ramshaw, J. D., & Chang, C. H. (1992). Computational fluid dynamics modeling of multicomponent thermal plasmas. *Plasma Chemistry and Plasma Processing*, *12*(3), 299–325. https://doi.org/10.1007/BF01447028

Ranjitha Gracy, T. K., Gupta, V., & Mahendran, R. (2019). Effect of plasma activated water (PAW) on chlorpyrifos reduction in tomatoes. *International Journal of Chemical Studies*, *7*(3), 5000–5006.

Rong, S. P., Sun, Y. B., & Zhao, Z. H. (2014). Degradation of sulfadiazine antibiotics by water falling film dielectric barrier discharge. *Chinese Chemical Letters*, *25*(1), 187–192. https://doi.org/10.1016/j.cclet.2013.11.003

Rong, S., & Sun, Y. (2014). Wetted-wall corona discharge induced degradation of sulfadiazine antibiotics in aqueous solution. *Journal of Chemical Technology and Biotechnology*, *89*(9), 1351–1359. https://doi.org/10.1002/jctb.4211

Sarangapani, C., Ziuzina, D., Behan, P., Boehm, D., Gilmore, B. F., Cullen, P. J., & Bourke, P. (2019). Degradation kinetics of cold plasma-treated antibiotics and their antimicrobial activity. *Scientific Reports*, *9*(1), 1–15. https://doi.org/10.1038/s41598-019-40352-9

Saravanan, A., Senthil Kumar, P., Jeevanantham, S., Karishma, S., Tajsabreen, B., Yaashikaa, P. R., & Reshma, B. (2021). Effective water/wastewater treatment methodologies for toxic pollutants removal: Processes and applications towards sustainable development. *Chemosphere*, *280*, 130595. https://doi.org/10.1016/j.chemosphere.2021.130595

Schwarzenbach, R. P., Egli, T., Hofstetter, T. B., Von Gunten, U., & Wehrli, B. (2010). Global water pollution and human health. *Annual Review of Environment and Resources*, *35*, 109–136. https://doi.org/10.1146/annurev-environ-100809-125342

Shen, R., Jiao, Z., Parker, T., Sun, Y., & Wang, Q. (2020). Recent application of computational fluid dynamics (CFD) in process safety and loss prevention: A review. *Journal of Loss Prevention in the Process Industries*, *67*, 104252. https://doi.org/10.1016/j.jlp.2020.104252

Sun, B., Aye, N. N., Gao, Z., Lv, D., Zhu, X., & Sato, M. (2012). Characteristics of gas-liquid pulsed discharge plasma reactor and dye decoloration efficiency. *Journal of Environmental Sciences*, *24*(5), 840–845. https://doi.org/10.1016/S1001-0742(11)60837-1

Sun, Y., Liu, Y., Li, R., Xue, G., & Ognier, S. (2016). Degradation of reactive blue 19 by needle-plate non-thermal plasma in different gas atmospheres: Kinetics and responsible active species study assisted by CFD calculations. *Chemosphere*, *155*, 243–249. https://doi.org/10.1016/j.chemosphere.2016.04.026

Taghvaei, H., Kondeti, V. S. S. K., & Bruggeman, P. J. (2019). Decomposition of crystal violet by an atmospheric pressure rf plasma jet: The role of radicals, ozone, near-interfacial reactions and convective transport. *Plasma Chemistry and Plasma Processing, 39*(4), 729–749. https://doi.org/10.1007/s11090-019-09965-w

Vörösmarty, C. J., McIntyre, P. B., Gessner, M. O., Dudgeon, D., Prusevich, A., Green, P., Glidden, S., Bunn, S. E., Sullivan, C. A., Liermann, C. R., & Davies, P. M. (2010). Global threats to human water security and river biodiversity. *Nature, 467*(7315), 555–561. https://doi.org/10.1038/nature09440

Wang, B., Dong, B., Xu, M., Chi, C., & Wang, C. (2017). Degradation of methylene blue using double-chamber dielectric barrier discharge reactor under different carrier gases. *Chemical Engineering Science, 168*, 90–100. https://doi.org/10.1016/j.ces.2017.04.027

Yu, S.-T., Chang, S.-C., Jorgenson, P., Park, S.-J., & Lai, M.-C. (1998). Basic equations of chemically reactive flows for computational fluid dynamics. *36th AIAA Aerospace Sciences Meeting and Exhibit*, 1051. https://doi.org/10.2514/6.1998-1051

Zeghioud, H., Nguyen-Tri, P., Khezami, L., Amrane, A., & Assadi, A. A. (2020). Review on discharge plasma for water treatment: Mechanism, reactor geometries, active species and combined processes. *Journal of Water Process Engineering, 38*, 101664. https://doi.org/10.1016/j.jwpe.2020.101664

Zeng, J., Yang, B., Wang, X., Li, Z., Zhang, X., & Lei, L. (2015). Degradation of pharmaceutical contaminant ibuprofen in aqueous solution by cylindrical wetted-wall corona discharge. *Chemical Engineering Journal, 267*, 282–288. https://doi.org/https://doi.org/10.1016/j.cej.2015.01.030

Zhang, T., Zhou, R., Wang, P., Mai-Prochnow, A., McConchie, R., Li, W., Zhou, R., Thompson, E. W., Ostrikov, K. (Ken), & Cullen, P. J. (2021). Degradation of cefixime antibiotic in water by atmospheric plasma bubbles: Performance, degradation pathways and toxicity evaluation. *Chemical Engineering Journal, 421*, 127730. https://doi.org/10.1016/j.cej.2020.127730

9 Filtration Process

*Raju Yerolla, P Suhailam, Praveen Kumar Ghodke, and Chandra Shekar Besta**
Department of Chemical Engineering, National Institute of
Technology Calicut, Kozhikode, Kerala–673601, India
*Corresponding Author: schandra@nitc.ac.in

CONTENTS

9.1 INTRODUCTION

Agriculture, the chemical industry, and environmental engineering use filtration frequently. Numerous filter processes can be identified based on the filter material, process parameters, or operating pressures, but they all utilize the same underlying principle: pouring a heterogeneous mixture through a filter membrane containing a predetermined number of pores. The barrier blocks larger particles while allowing smaller particles to pass. The fluid may pass through the membrane, but it will experience difficulty to flow due to particle buildup and pore restriction. In one of the earliest attempts to replicate filtering processes using CFD-DEM, Li and Marshall [1] modeled a cylindrical fiber in an array for microparticle deposition, accounting for the adhesive elastic interactions between the fiber and particle. Due to particle shadowing and strong shear stresses, it was discovered that particles tend to settle along the fiber front toward the center in a confined zone.

Following particle deposition, the model does not account for changes in porosity in the fibrous medium and flow field. Except for the drag force, which was approximated using the Di Felice drag equation, no fluid forces were considered. Qian et al. [2] created a model for modeling gas-flow characteristics inside a fibrous medium and investigated the effect of fiber structure and particle quality on deposition and agglomeration features in the filtering process. The fibrous media were modeled in three dimensions using scanning electron microscope data. They discovered that the empirical correlation values and model filtering efficiency were consistent. Qian et al. [3] used CFD-DEM to properly characterize the interactions between filter material and particles in a fibrous medium exposed to particle loading. Gas–solid flow properties in particle-loaded fibrous medium were predicted using the Hertz–Mindlin contact model in the JKR-cohesion model variant. Two-way coupling was used to account for particle–particle, particle–fluid, and particle–filter material interactions. The deposition mechanism of submicron particles interacting with a single cylindrical fiber was examined quantitatively by Dong et al. [4]. They observed two steps of deposition: particle deposition on the fiber surface and subsequent agglomeration development into dendritic branches.

Using CFD-DEM, Naukkarinen et al. [5] investigated the hydrodynamics of a mixed ion-exchange membrane filtering unit for wastewater treatment. As the quantity of particles deposited during pressure filtering rose, so did the cake's thickness [1]. Their CFD-DEM model accounted for both solid and liquid properties, including compressibility, polydispersity, density, and viscosity. By comparing the produced cake resistance and filtrate flow data to experimental data, the model was effectively verified. Due to the often small size of the particles to be described, only the microscale, which consists of tiny holes and fibers, is typically represented, and macroscopic filter coefficients are calculated. These may then be utilized to construct larger-scale simulations of genuine filtering devices. However, the present state of the art in filtering unit design has not fully realized the benefits of more complex strategies that may include a broad range of characteristics, such as the work of Li et al. [1]. Although the CFD-DEM approach may be deemed mature for this purpose in terms of physical models, its commercial application cannot be ascribed to inadequate development of multi-scale methods.

Filtration with cake formation is a very efficient technique for removing particles from a suspension. After filtering, it is possible to consolidate the cake to reduce its moisture content. This is often accomplished using so-called press-filters [6]. The cake is composed of small particles whose compressibility varies based on the particle contact stiffness and packing structure, which are the primary determinants of compressibility. Following filtration, it is preferable to apply mechanical pressure to the cake in this second phase, resulting in a decrease in cake height (h_c). Due to the cake's shrinking volume, the amount of water in the pores is reduced to a bare minimum. This results in the formation of an extremely dense liquid-saturated particle packing.

During the filtration process, the liquid goes through a compressed particle packing, and the differential flow rate between the solid and liquid must be considered [7]. The particles cluster together to form micron-sized aggregates. In this particle size range, solid-liquid separation is governed more by the interfacial effects of aggregates than by the size of the primary particles [8], and the size of the primary particles has limited impact. Between particles, electrostatic repellent and Van der Waals attractive interactions produce flocculation. This may lead to the production of flocs and the separation of particles. In the scientific community, the DLVO-theory is a well-known physical model of colloid stability [9, 10]. The liquid drag force between the particles is produced by the mechanics of cake manufacturing, which are described in further detail below. Consequently, axial particle pressure (P_s) is also conveyed in the packing connections. The resistance of the filter cake increases with filtration length and cake height, resulting in a slower infiltration rate. In the micron range, the ratio of flow during the manufacturing and compression of ultrafine porous particle packings may be examined by integrating discrete element modeling with fluid dynamics.

In the DEM software, a "fixed course-grid fluid scheme" is used to establish the fluid link. According to [11], the system provides solutions for the locally averaged, two-phase mass and momentum equations for pressures and fluid velocities; this may be seen as a generalized version of the Navier–Stokes equation for a solid phase interacting fluid. The fluid solver employs the SIMPLE method [12] to solve incompressible viscous or inviscid flow on a predefined rectangular shape aligned with the Cartesian axes. Formally, the internal discretization is regular and consistent. The fundamental coupling concept posits that the particle's radius is negligible compared to the length of a single fluid element in the system. At any Reynolds number, there is no contact with barriers, and no turbulence factors are added into the equation.

The fluid connection is achieved using a "fixed course-grid fluid scheme" implemented in the DEM software. The system, according to [11], gives solutions to the locally averaged, two-phase mass and momentum equations for pressures and fluid velocities, this may be thought of as a generalized variant of the Navier–Stokes equation for a solid phase interacting fluid. The fluid solver uses the SIMPLE approach [12] to solve inviscid flow or incompressible viscous on a fixed rectangular geometry aligned with the Cartesian axes in incompressible viscous or inviscid flow. In terms of form, the internal discretization is regular and fixed. The underlying coupling notion assumes that the radius of the particle is insignificant in comparison to the length of a single fluid element in the system. There is no contact with obstacles at any Reynolds number, and no turbulence variables are introduced to the equation.

To inhibit fluid flow, the boundary conditions should be placed on the outer barrier of the fixed rectangular form. Among the circumstances that may be employed are fixed pressure or fixed normal velocity, as well as fixed or free shear velocity. Tsuji conducted pioneering research on particle/fluid interaction [13, 14], which was also noted in [15]. There are no other mechanisms for the generation of pore pressure while under strain, according to the assumptions of this approach. In the discrete element method software, the solution is referred to as "pseudo-3D." While the two dimensions are solved by the Navier–Stokes equation, the porosity of the fluid is defined in terms of the total volume of spheres in the fluid. This is because when compared to 3D measurements of the identical items, the two-dimensional area of the circles would overestimate the particle forces and underestimate the porosity. The cells are assumed to have an out-of-plane length equal to the largest particle diameter in the model in porosity calculations, and the particles are assumed to be spheres. To maintain the pseudo-3D geometry, the particle volume with diameters less than the maximum diameter of the particle is scaled by $(d_{max}/d_{particle})$. For example, actual particle investigating the impact of agglomeration, generated by DLVO interactions, on the production of filter cakes are available [15, 16]. A mixed computing strategy is advocated by the authors, in which the cake production process is simulated using both a stochastic rotation dynamic (SRD) and a molecular dynamics method. Keller et al. [17] performed a simulation (numerical) research superimposed grid based approach that encompasses the many microscopic effects of the double layer interaction for a small particle number. They discovered the association between permeability and particle size and porosity, as well as the relationship between permeability and particle agglomeration.

When particle counts are higher, a combination of DEM and SRD is utilized to depict cake formation and agglomeration. Peng et al. [16] created a DEM-based model for particle transport in colloidal suspensions that takes into account both Brownian motion and the external movement of particles caused by electrophoresis as particle transport driving variables. Using the DLVO theory, they investigated the influence of volume concentration and pH value of Alumina agglomerates in water on formation. Dong et al. [4] investigated the formation and development of filter cakes using DEM. Van der Waals attraction accounted for and modeled the influence of particle and fluid material qualities as well as external factors.

9.2　CFD-DEM MODEL FORMULATION

9.2.1　DEM-DISCRETE ELEMENT METHOD

The discrete element technique was developed by Cundall [18] for the study of rock mechanics issues, and it was extended to soils by Cundall and Struck [19]. The model consists of separate particles, which move independently of each other and only interact at interfaces or contacts. Because the particles are stiff, the contact behavior is explained by using the approach of soft contact. A normal stiffness is measure stiffness at a contact, and the mechanical behavior of such a system is defined in terms of the movement of each particle and the inter-particle forces acting at each contact site. The laws of motion define the fundamental relationships between particle motion and the forces that create that motion. In discrete element method simulation, a particle has rotational and translational motion, which may be specified by particle velocity v_i or particle angular velocity ω_i, the mass m of a particle m_i, the momentum T_i and force

$$m_i \frac{dv_i}{dt} = F_i \tag{9.1}$$

$$I_i \frac{d\omega_i}{dt} = T_i \tag{9.2}$$

The contact model proposed by Hertz and Mindlin is frequently used to describe particle-wall or particle-particle interactions. It is made up of many well-known material parameters and describes the tangential and normal force acting on a particle in the contact zone. Equation (9.3) may be used to compute the normal force, here $E_{M,ij}$ is the average Young's modulus of particles i and j, $r_{P,M}$ is the average particle radius, and s_N is the ovelap of the particles vertical to the contact surface.

$$F_N = \frac{4}{3} E_{M,ij} \sqrt{r_{P,M} s_N} \tag{9.3}$$

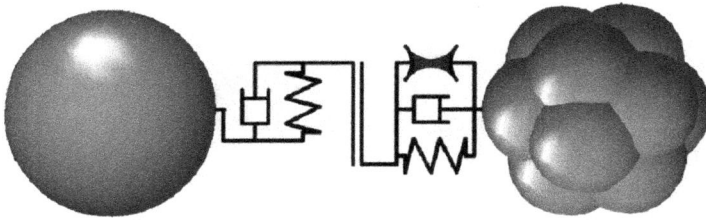

FIGURE 9.1 Contact law. Copyright © CC BY 3.0 [20].

The damping force $F_{N,D}$, which takes into account energy dissipation, is determined by the coefficient of restitution, particle mass m_p, and the relative velocity of the contact point in the normal direction. The Coulomb law takes into consideration fully formed friction between two particles.

$$F_C = \mu_{\text{fric},ij} F_N \qquad (9.4)$$

The PFC (Particle Flow Code) software is used to compute the contact force in the normal direction based on a parallel connection of a dashpot element and spring – and a repulsive or adhesive force as a result of the particle's surface charges. The tangential contact force is caused by the parallel connection of a spring and dashpot element, as well as their series 30 connection with a friction element. This configuration is depicted visually in Figure 9.1.

The resultant contact force in normal direction $F_{k,N}$ is the total of the damping force F_{damp}, spring force F_{spring}, and repulsive or adhesive force due to the surface charge F_{surface}.

$$F_{k,N} = F_{\text{spring}} + F_{\text{damp}} + F_{\text{surface}} \qquad (9.5)$$

where

$$F_{\text{spring}} = k_n^{ij} s_n^{ij} \qquad (9.6)$$

$$F_{\text{damp}} = \eta_n^{ij} v_n^{ij} \qquad (9.7)$$

$$F_{\text{surface}} = F_{\text{adhesive}} + F_{\text{repulsive}} \qquad (9.8)$$

The Van der Waals force between two particles is represented by the adhesive force, which may be computed using the Hamaker constant C_H.

$$F_{\text{adhesive}} = -\frac{C_H r}{12a^2} \qquad (9.9)$$

The repulsive force varies with particle radius r, Avogadro constant N_A, absolute temperature T, Boltzmann constant K_B, electrolyte concentration c, Debye Hückel parameter κ, and Zeta-potential ς.

$$F_{\text{repulsive}} = \frac{67\pi r c N_A \Gamma^2 K_B T}{\kappa} \qquad (9.10)$$

$$\Gamma = \tanh\left(\frac{ze\varsigma}{4K_B T}\right) \qquad (9.11)$$

From equations (9.8–9.11), it is noted that the surface force may be defined as the product of a particular surface energy, which includes the previously given factors, and the particle diameter d. In addition to the contact rule, the crucial step time is required for a stable simulation. The oscillation of a massless spring of one dimension with stiffness k and a linked mass point of mass m yields the crucial time step. This leads to the translation's essential step time.

$$\Delta t_{\text{crit,tran}} = \sqrt{\frac{m}{k_{\text{tran}}}} \qquad (9.12)$$

According to equation (9.12), the critical time step for ultrafine particles ($d<10\mu m$) may reach values as high as $\Delta t < 10^{-9}\text{s}$. Because of the limited capacity of CPU, real-time simulations of such systems are difficult. As a result, the size of model is substantially less than the experimental size.

9.2.2 CFD AND DEM COUPLED

To simulate the flocculation of ultrafine particle suspensions, an attractive force based on the DLVO theory [9, 10] was created. The fixed coarse-grid fluid flow technique is used in the DEM application to combine the discrete element method with fluid dynamics. The system, as previously stated, solves the locally averaged, two-phase momentum and mass equations for fluid velocities and pressures, which may be thought of as a generalized form of the Navier–Stokes equation for a fluid interacting with a solid phase. To account for the influence of a particle solid phase integrated into the fluid, the Navier–Stokes equations for laminar flow or incompressible viscous flow may be modified. Porosity may be used to characterize the average impact of a large number of particles [11]. Figure 9.2 depicts the flowchart of the CFD-DEM coupling.

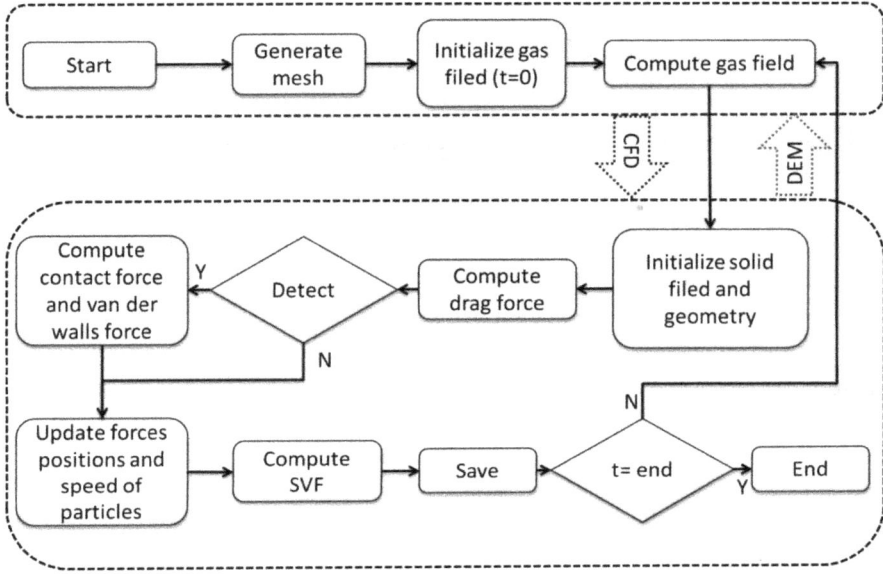

FIGURE 9.2   CFD–DEM coupling flowchart.

$$\rho_f \frac{\partial \overrightarrow{\varepsilon u}}{\partial t} + \rho_f \vec{v}.\nabla\left(\overrightarrow{\varepsilon u}\right) = -\varepsilon\nabla p + \mu\nabla^2\left(\varepsilon\vec{u}\right) + \overrightarrow{f_b} \qquad (9.13)$$

$$\frac{\partial \varepsilon}{\partial t} + \nabla.\left(\varepsilon\vec{u}\right) = 0 \qquad (9.14)$$

The drag force per unit volume is denoted by f_b. The fluid velocity is indicated by the symbol $\vec{u}$. The drag force of a fluid flow around a particle is denoted by

$$\overrightarrow{f_b} = \beta\left(\vec{u} - \vec{v}\right) \qquad (9.15)$$

where $\vec{v}$ is the average velocity of all particles in a fluid element. The coefficient β is determined by the fluid element's porosity. The ERGUN connection is utilized for low porosity values ($\varepsilon < 0.8$) [21].

$$\beta = \frac{(1-\varepsilon)}{d_{50}^2 \varepsilon^2}\left(150(1-\varepsilon)\mu + 1.75\rho_f d_{50}|\vec{u} - \vec{v}|\right) \text{ for } \varepsilon < 0.8 \qquad (9.16)$$

For larger values of solid percent ($\varphi_s > 0.8$), β is calculated from the corrected non-linear drag force exerted by a fluid on a spherical particle [22].

$$\beta = \frac{4}{3} C_d \frac{|\vec{u} - \vec{v}| \rho_f (1-\varepsilon)}{d_{50} \varepsilon^{1.7}} \quad \text{for } \varepsilon > 0.8 \tag{9.17}$$

C_d is the turbulent drag coefficient defined in terms of the particle Reynolds number.

$$C_d = \begin{cases} \dfrac{24\left(1 + 0.15 Re_p^{0.687}\right)}{Re_p} & Re_p < 1000 \\ 0.44 & Re_p > 1000 \end{cases} \tag{9.18}$$

where

$$Re_p = \frac{|\vec{U}| \varepsilon \rho_f d_{50}}{\mu} \tag{9.19}$$

And $|\vec{U}| = |\vec{u} - \vec{v}|$ denotes the average relative velocity of the particles and fluid.

9.3 EXPERIMENT

Micro glass spheres with an average particle size of 5.8 μm were utilized in the filtering studies. Table 9.1 shows the attributes of the glass spheres. The suspensions were made using distilled water and a solid volume fraction of 13%. A chemical conditioning of the suspension is beneficial to enhance filtration behavior by changing the inter particular repulsive and attractive forces by adding electrolytes or modifying the pH-value.

To test the concept experimentally, material parameters such as packing density $\varepsilon_s, 0$ and compressibility index β, permeability k_0 at $Ps = 0$, lateral pressure ratio λ_w, and pure medium filter medium resistance $R_{FM,0}$ must be measured. The press-shear-cell method was used to calculate these parameters (Figure 9.3). In the medium pressure range, it combines a laboratory filter, a ring shear cell, and

TABLE 9.1
Particle Properties of Glass Spheres

Filtration pressure p in kPa	200–800
Particle diameter d_{50} in μm (primary particles)	5.8
Solid density ρ_s in kg/m³	2520
Particle stiffness k_n/k_s in N/m	1350
Friction coefficient	0.5
Zeta potential ζ in mV	-40 (stabilized) ≈0 (flocculated)
Hamaker constant, $C_{H,sls}$ in 10^{-20} J	0.43

Source: Copyright © CC BY 3.0 [20]

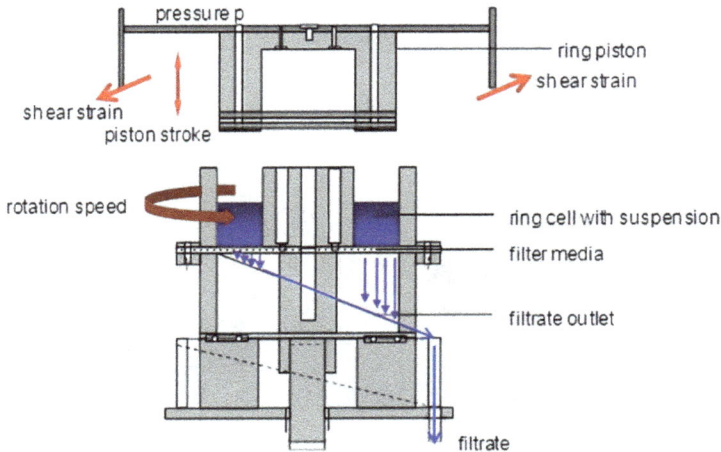

FIGURE 9.3   Pressure shear cell. Copyright © CC BY 3.0 [20].

a compression-permeability cell. This test instrument is best suited for use in continuous pressure filtering operations.

9.4   RESULTS AND DISCUSSION

For the initial configuration, particles are randomly distributed in the representative volume element with a beginning velocity of $v = 0$ and no particle overlapping. During the simulation, the filtration pressure $p = 200$ kPa. The particle number is changed depending on the size of model to achieve a solid volume percentage of 13 percent. Using the Darcy equation, the volume flow rate of the filtrate is computed at each step time. As a result, the time-dependent filtrate volume for each step time may be computed. If all the particles are settled for the particular pressure, then the simulation is over. To guarantee that all particles are filtered, the filter medium at the bottom is mimicked using point walls with a maximum distance of 0.75 times the minimum particle diameter inside the representative volume element. Figure 9.4 depicts the DEM simulation's beginning condition.

When substantial repulsive interactions of suspension dominate, the particles become totally deposited and dispersed as single particles. These suspensions result in a dense filter cake (Figure 9.5a) with high resistances and low permeability.

The flocculation achieves its greatest value only if the attractive forces exist between the ultrafine particles in the suspension. Large floes deposit quickly, resulting in a loose, unstable structure with a high compressibility (Figure 9.5b). The filtering time might be reduced due to flocculation. For the two circumstances, the porosity time-dependent curves of inside the lowest fluid cell were recorded and compared. Figure 9.6 depicts how the cake porosity reduced rapidly at first and gradually neared equilibrium levels. The equilibrium state is attained more quickly in the case of a flocculated suspension, which has a substantially larger porosity.

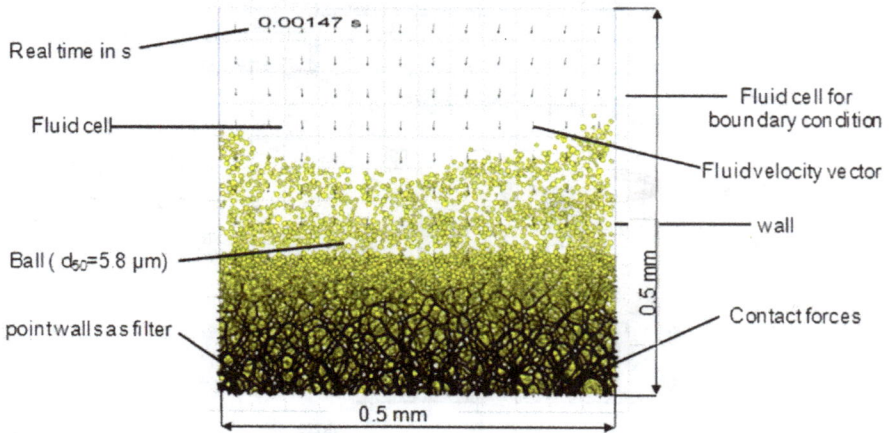

FIGURE 9.4 Initial state of DEM simulations. Copyright © CC BY 3.0 [20].

The compression function was used to modify the packing density ε_s for various filtration pressures p_s in order to forecast how applied pressure depends on the filter cake structure.

$$\varepsilon_s(p) = \varepsilon_{s,0}\left(1+\frac{p_s}{p_0}\right)^{\beta} \tag{9.20}$$

where β represents the compressibility index and p_0 represents the compression modulus.

Figure 9.7 shows that higher pressures result in more compressed filter cakes. The flocculated cake is classified as extremely compressible and the compressibility is greater than that of the stable suspension. Quantitatively, the simulation findings accord with the experimental values.

9.5 CONCLUSION

The filtration dynamics of ultrafine particle suspensions were characterized and simulated in this study using a coupled DEM and CFD. The numerical outcomes were compared to the experimental findings. To explore the compression nature of the consolidated particle packing and to measure the most essential material parameters, a press-shear-cell method was utilized. This approach was used to estimate the

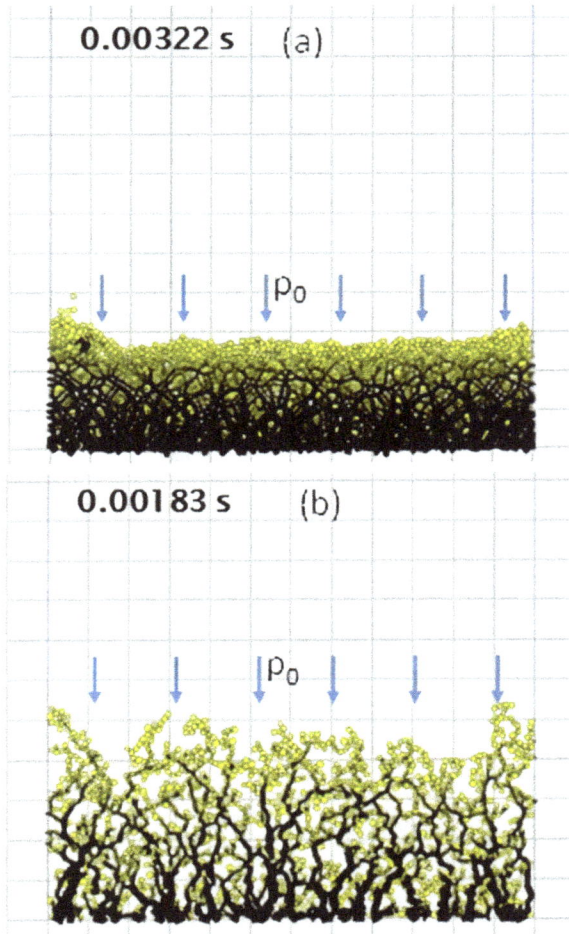

FIGURE 9.5 Simulation result (a) Stable suspension and (b) flocculated suspension. Copyright © CC BY 3.0 [20].

time dependence of cake height and specific filtrate volume for non-flocculated and flocculated suspensions. It is found that flocculation increases the permeability and porosity of the filter cake while decreasing the filtering time. A stable suspension with dominating repulsion results in a compact filter cake with low porosity, high filter cake resistance, and low permeability.

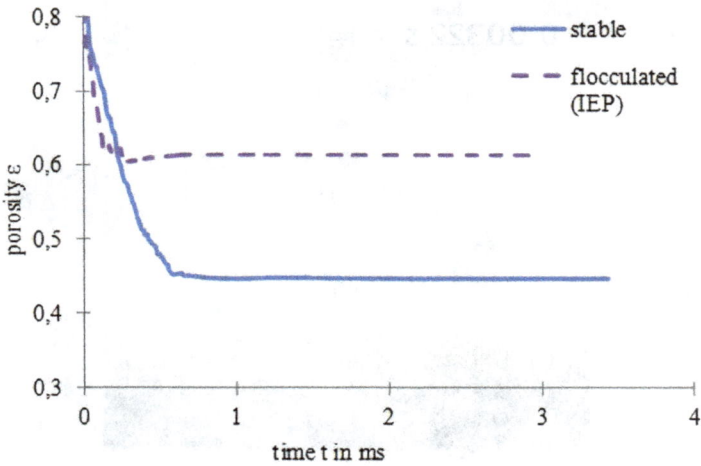

FIGURE 9.6   Porosity of the lowest fluid cell versus time. Copyright © CC BY 3.0 [20].

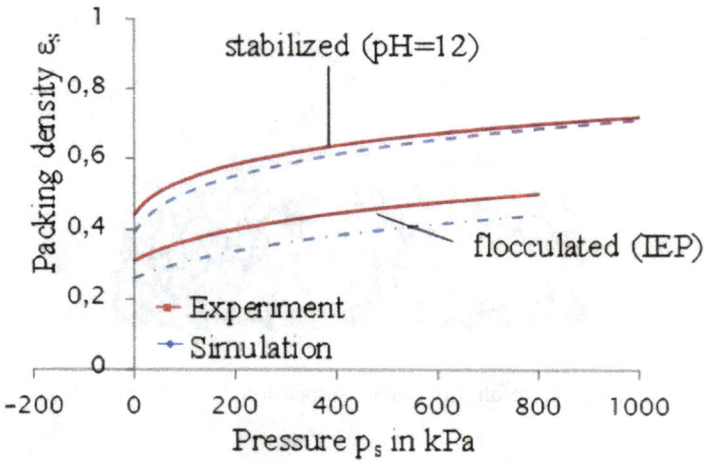

FIGURE 9.7   Packing density versus filtration pressure. Copyright © CC BY 3.0 [20].

REFERENCES

[1] S.-Q. Li and J. S. Marshall, "Discrete element simulation of micro-particle deposition on a cylindrical fiber in an array," *J. Aerosol Sci.*, vol. 38, no. 10, pp. 1031–1046, Oct. 2007, doi: 10.1016/j.jaerosci.2007.08.004.

[2] F. Qian, N. Huang, X. Zhu, and J. Lu, "Numerical study of the gas–solid flow characteristic of fibrous media based on SEM using CFD–DEM," *Powder Technol.*, vol. 249, pp. 63–70, Nov. 2013, doi: 10.1016/j.powtec.2013.07.030.

[3] F. Qian, N. Huang, J. Lu, and Y. Han, "CFD–DEM simulation of the filtration performance for fibrous media based on the mimic structure," *Comput. Chem. Eng.*, vol. 71, pp. 478–488, Dec. 2014, doi: 10.1016/j.compchemeng.2014.09.018.

[4] K. J. Dong, R. P. Zou, R. Y. Yang, A. B. Yu, and G. Roach, "DEM simulation of cake formation in sedimentation and filtration," *Miner. Eng.*, vol. 22, no. 11, pp. 921–930, Oct. 2009, doi: 10.1016/j.mineng.2009.03.018.

[5] T. Naukkarinen, M. Nikku, and T. Turunen-Saaresti, "CFD-DEM simulations of hydrodynamics of combined ion exchange-membrane filtration," *Chem. Eng. Sci.*, vol. 208, p. 115151, Nov. 2019, doi: 10.1016/j.ces.2019.08.009.

[6] C. Alt, "Schlammentwässerung mit Preßfiltern," *Chem. Ing. Tech.*, vol. 48, no. 2, pp. 115–124, 1976, doi: 10.1002/cite.330480209.

[7] B. Reichmann, "Modellierung der Filtrations-und Konsolidierungsdynamik beim Auspressen feindisperser Partikelsysteme," PhD Thesis, Verlag nicht ermittelbar, 1999.

[8] W. Gösele, "Grenzflächenkräfte und Fest-Flüssig-Trennung-was der Praktiker davon wissen sollte (Interfacial effects and solid/liquid-separation—what the practitioners should know)," *Filtr. Sep.*, vol. 9, pp. 14–22, 1995.

[9] E. J. W. Verwey, J. T. G. Overbeek, and K. van Nes, "Theory of the Stability of Lyophobic Colloids: The Interaction of Sol Particles Having an Electric Double Layer," p. 205, 1948.

[10] B. Derjaguin and L. Landau, "Theory of the stability of strongly charged lyophobic sols and of the adhesion of strongly charged particles in solutions of electrolytes," *Prog. Surf. Sci.*, vol. 43, no. 1, pp. 30–59, May 1993, doi: 10.1016/0079-6816(93)90013-L.

[11] J. X. Bouillard, R. W. Lyczkowski, and D. Gidaspow, "Porosity distributions in a fluidized bed with an immersed obstacle," *AIChE J.*, vol. 35, no. 6, pp. 908–922, 1989, doi: 10.1002/aic.690350604.

[12] S. V. Patankar, *Numerical Heat Transfer and Fluid Flow*. CRC Press, 2018. doi: 10.1201/9781482234213.

[13] Y. Tsuji, T. Kawaguchi, and T. Tanaka, "Discrete particle simulation of two-dimensional fluidized bed," *Powder Technology*, vol. 77, no. 1, pp. 79–87, 1993, doi: 10.1016/0032-5910(93)85010-7.

[14] T. Kawaguchi, "Numerical simulation of fluidized bed using the discrete element method," *Jpn Soc Mech Eng Ser B*, vol. 58, pp. 79–85, 1992.

[15] Y. Shimizu, R. Hart, and P. Cundall, Numerical Modeling in Micromechanics via Particle Methods–*2004: Proceedings of the 2nd International PFC Symposium, Kyoto, Japan, 28–29 October 2004*. CRC Press, 2004.

[16] Z. Peng, Z. Yuan, K. Liang, and J. Cai, "Ice slurry formation in a cocurrent liquid-liquid flow," *Chin. J. Chem. Eng.*, vol. 16, no. 4, pp. 552–557, Jan. 2008, doi: 10.1016/S1004-9541(08)60120-2.

[17] F. Keller, C. Eichholz, B. Schäfer, and H. Nirschl, "Numerical sumulation of agglomeration and filtration of colloidal suspensions." in *FILTECH Proceedings*, 2011, vol. 1, no. s, pp. 275–292.

[18] P. A. Cundall, "A computer model for simulating progressive, large-scale movement in blocky rock system," *Proocedings of the Symposio of the International Society of Rock Mechanics, Nancy 2*, 1971, p. No. 8.

[19] P. A. Cundall and O. D. L. Strack, "A discrete numerical model for granular assemblies," *Géotechnique*, vol. 29, no. 1, pp. 47–65, Mar. 1979, doi: 10.1680/geot.1979.29.1.47.

[20] S. Sören and T. Jürgen, "Simulation of a filtration process by DEM and CFD," *Int. J. Mech. Eng. Mechatron.*, vol. 1, no. 2, pp. 28–35, 2012.

[21] S. Ergun, "Fluid flow through packed columns," *Chem Eng Prog*, vol. 48, pp. 89–94, 1952.

[22] C. Y. Wen and Y. H. Yu, "Mechanics of fluidization," *Chem Eng Prog Symp Ser*, vol. 62, pp. 100–111, 1966.

10 Water Treatment by Microbial Activity

Chingakham Ngotomba Singh, Niranjan Thota,
*E.J. Rifna, and Madhuresh Dwivedi**
Department of Food Process Engineering, National Institute
of Technology Rourkela, Rourkela, Odisha, India
*Corresponding Author

CONTENTS

10.1 INTRODUCTION

It is widely acknowledged that some climate change caused by humans cannot be stopped. There has been a lot of focus on potential effects on water supply, but very little is known about the corresponding changes in water quality. The forecasted variations in rainfall and air temperature may have an impact on river flows, which may in turn influence pollutant mobility and dilution. In conjunction with quality declines, rising water temperatures will have an impact on the kinetics of chemical reactions and the biological condition of freshwater. Increased flows would result in changes in stream power and, as a result, sediment loads. These changes might affect lake and stream systems' freshwater ecosystems by changing the shape of rivers and the way sediment is transported to lakes (Whitehead et al., 2009).

For carbon-based life, the kind of life we are familiar with, water is necessary. Although it is possible to imagine different compound-solvent pairings that may occur in very particular physical circumstances, the components required for carbon- and water-based life are among the most prevalent in the universe (Westall & Brack, 2018). It is stated that water is the "matrix of life" and that life cannot exist without it (Ball, 2017). Life needs organic and inorganic matter to function the metabolite. The earliest complex organic molecules, and eventually the first living forms, most likely developed in an ancient sea that included minerals that gave the substances structural stability and catalytic capacity (Nielsen, 2000). But also water quality is declining due to the overload of organic matter and minerals, and there are more scenarios where it might have an adverse impact on both plants and human health (Delpla et al., 2009). Water treatment is employed to maintain the water quality and prevent harmful outcomes from impurities.

Water treatment is a procedure that involves many operations (physical, chemical, and biological) to remove and reduce pollution or undesirable properties of water. The different components, such as organometallic substances, minerals, heavy metals, and many other organisms, act as contaminants when interacting with water and make them inaccessible. Some less appealing impurities include oil and scum, organic matter, fish, boards, rags, and anything that may be dumped into the sewage system to the waterways.

10.2 FACTORS AFFECTING THE QUALITY OF WATER

The water quality of numerous rivers across the world is deteriorating, as evidenced by, for instance, changing amounts of nutrients, salts, and sediments. Understanding how and why water quality varies over time, both within and between river catchments, is essential for effective management of water quality. The main components of a catchment that have an impact include its topography, climate, geology, usage of land, atmospheric deposition, and hydrology of the catchment are as follows (Jachimowski, 2017; Lintern et al., 2018):

(a) The matter sources
   • The organic or inorganic matter that can be runoff to the water body through river or any stream ways
   • Granular size and structure
   • Granular stability
   • Dissolved oxygen concentration
   • Presents of inoculum
(b) Mobilization
   • Temperature
   • Water stream velocity
   • Presence of barrier in flow
(c) By distribution
   • Efficiency of treated water distributed in the water supply systems
   • Decontamination of by-products
   • Insufficient water biostability and resulting microbiological growth
   • Lack of water biostability, which causes it to be corrosive and accumulate sediment
   • Pressure alterations
   • Water flow rate
   • Water stopped to flow and stagnating in the network

10.3   ROLE OF MICROORGANISMS IN WATER TREATMENT

Microorganisms have been shown to be an effective, affordable, and environmentally acceptable alternative to physicochemical approaches and play a crucial part in the bioremediation process. The ability of diverse microbial species, including fungus, bacteria, and algae, to break down and discolor harmful chemical contaminants found in various industrial effluent sources, including distilleries, has been examined. Numerous studies have been conducted on free or immobilized cells for distillery wastewater bioremediation (Bezuneh, 2016). The nature of chemical composition of wastewater, nutrients, pH, temperature, oxygen, and inoculum size are all important factors in determining the potential of microorganisms in water treatment (Gay et al., 1996; Shahid et al., 2020).

Most modern treatment facilities utilize aerobic bacteria in a condition known as an aerated environment. The contaminants in the water are broken down by this bacteria using the free oxygen present in the water, which it then transforms into energy for growth and reproduction (Craggs et al., 2013). Normally, anaerobic microorganisms are utilized in wastewater treatment. These bacteria's primary feature in the treatment of sewage is to minimize the amount of sludge and turn it into methane gas. Anaerobic microorganisms utilized in sewage treatment also have the benefit of removing phosphorus from wastewater (Show & Lee, 2017).

In sewage treatment, facultative microorganisms are bacteria that can switch between anaerobic and aerobic states depending on their surroundings. These bacteria typically prefer an aerobic environment. Particularly in the winter, the facultative and aerobic bacteria might use oxygen to convert more ammonia to nitrate (nitrification), breakdown more organic matter, and produce more nitrate (Tao et al., 2010).

10.4   IMPORTANT MICROBES IN WATER TREATMENT

The employment of microorganisms in water treatment eliminates biodegradable complicated hazardous chemicals into acceptable and acceptable end products, such as CO_2 and H_2O (Table 10.1). Biofilm and biological floc are formed by trapping the suspended and non-settling colloid. It is also possible to separate and recover metals, certain organic components, and significant nutrients (Rani et al., 2019).

There are several varieties of microorganism that are taken in the water treatment process because a single bacteria cannot accomplish the treatment effectively. It needs a combination of different types of bacteria for efficient treatment such as nitrifying bacteria, denitrifying bacteria, and methanogens bacteria (Ochoa-Herrera et al., 2009). In suspended growth cultures, protozoa are crucial because they graze on colloidal organic matter and bacteria that are dispersed, which reduces the amount of turbidity that is left over after the biofloc has been removed by sedimentation (Grady Jr et al., 2011).

Nitrifying bacteria are aerobic bacteria and they can convert nitrogenous waste into a nitrate. Ammonia has high BOD because NH_3 oxidation requires oxygen. This nitrogenous ammonia is toxic to health but when it is converted into nitrate, it becomes less toxic. There are two groups of microbes involved such as AOB (ammonia-oxidizing bacteria) and NOB (nitrite-oxidizing bacteria). When compared to ammonia oxidizers, nitrite-oxidizing bacteria have more phylogenetical diversity.

The majority of the denitrifying bacteria are anaerobic. Nitrate may be changed into nitrogen by them, and the nitrogen can then be released into the atmosphere or environment. Some *Serratia, Pseudomonas,* and *Achromobacter* species as well as *Thiobacillus denitrificans* and *Micrococcus denitrificans* have been linked to denitrification.

The methanogens bacteria can deliver methane gas that is converted from the organic substance. They develop indirectly by dissolving the existing polymers into monomers rather than directly interacting with the organic constituents. Following that, the fermentation of those monomers yields CO_2 and acetate. The next step is the production of methane gas from the CO_2 and acetate. Methanogens come in three different types: CO_2 reducing, methyl reducing, and acetoclastic.

TABLE 10.1
Several Microorganisms of Ammonia-Oxidizing Bacteria and Nitrite-Oxidizing Bacteria

AOB (ammonia-oxidizing bacteria)	Organisms	Functions
	Nitrosomonas spp.	• Ammonia is oxidized to nitrite • Increase the bioavailability of nitrogen to plants and in the denitrification • Common nitrifiers used in aquaculture • β-Proteobacteria
	Nitrosococcus spp.	• Oxidize ammonia to nitrous acid • Common nitrifiers used in aquaculture
	Nitrosospira spp.	• Ammonia is oxidized to nitrite • β-Proteobacteria
	Nitrosolobus spp.	• Ammonia is oxidized to nitrite
	Nitrosovibrio spp.	• Common nitrifiers used in aquaculture
NOB (nitrite-oxidizing bacteria)	*Nitrobacter* spp.	• Oxidize nitrite to nitrate • α-proteobacteria • Possess intracytoplasmic membranes
	Nitrococcus spp.	• γ-Proteobacteria • Possess intracytoplasmic membranes
	Nitrospira spp.	• δ-proteobacteria
	Nitrospina spp.	• Forming its own subdivision • Lacks intracytoplasmic membranes

(Braker & Conrad, 2011; Jasmin et al., 2020; Samocha, 2019)

10.5 MECHANISM IN WATER TREATMENT

10.5.1 PHYSICAL TREATMENT

In all water treatment, physical treatment is carried out initially. The preliminary and primary treatment are categorized under physical treatment. There are several steps involved in the physical treatment of water such as screening, settling, flotation, filtration, and centrifugation.

10.5.1.1 Screening

Screening is the initial step in the treatment of water and waste. Screening nets or bars are used to remove large particles while safeguarding downstream components like filters, pumps, and pipelines. Wood, plastic, and paper are examples of larger particles that are separated using perforated film.

10.5.1.2 Settling

The aggregation of solid particles referred to as iron sulphate is an example of a suspended solid whose density is higher than that of water, and which settles to the bottom surface. The three most important variables that impact the diameter, density, and viscosity of the solution are those that affect how quickly the SS settles. How multiple components relate to one another is described by Stokes' equation. Following coagulation and flocculation treatment employing coagulants and flocculants, small-scale solids and colloids that are hardly settled naturally are settled.

10.5.1.3 Flotation

Oils and fats are separated from the water via flotation because of their densities, which are similar to or less than that of water. Stokes' equation also includes a term for the flotation velocity of particles. As a result, under natural circumstances, particles with big sizes and low density can float with ease. Mechanical flotation and dissolved air flotation are employed to accelerate the floating velocity of particles.

10.5.1.4 Filtration

The filtration may be achieved by filtering water through screen, bag, cartridge or similar filters, multi-media depth filters, and membrane filters. Using the different filters, there are six methods of physical water treatment such as greensand filtration, multimedia filtration, microfiltration, ultrafiltration, nanofiltration, and reverse osmosis.

10.5.1.4.1 Greensand filtration

It is made of glauconite that has been coated with manganese oxide. It is perfect for filtering dissolved manganese, hydrogen sulphide, and iron. Generally, it has to be pre-treated with an oxidizing agent.

10. 5.1.4.2   Multimedia filtration

It contains a minimum of three separate filtering layers. It was once regarded as the ideal filter for particles and suspended solids more than 10–25 microns. It does not get rid of germs, viruses, or microscopic protozoa.

10.5.1.4.3   Microfiltration

The barrier membrane is used to get rid of suspended particles. Filtering particles as small as 0.1–10 microns is ideal (e.g., algae, protozoans). Contaminants that have been dissolved are not removed. It can be used to lower suspended particles to safeguard reverse osmosis membranes.

10.5.1.4.4   Ultrafiltration

Suspended solids are removed using the barrier membrane. Filtering down to microns is ideal (e.g., bacteria and some viruses). Additionally, it does not get rid of dissolved impurities. On water with a high concentration of microscopic suspended particles, pre-treatment is frequently used before RO systems. The membranes are protected by lowering the silt density index (SDI).

10.5.1.4.5   Nanofiltration

The semi-permeable membrane is utilized to filter out certain dissolved solids and suspended particles. It is perfect for removing particles as small as 0.001 microns (e.g., bacteria, viruses, divalent, and trivalent ions)

10.5.1.4.6   Reverse osmosis

Suspended solids are removed using a semi-permeable membrane. Contaminants as small as 0.0001 microns can be filtered out using it (e.g., dissolved ions, bacteria, and viruses). With a single pass, it may remove up to 98% of the dissolved ions, making it one of the most economical deionization processes.

10.5.1.5   Centrifugation

The settling of particles in a liquid-solid emulsion is facilitated by the centrifugation separation process. This method is frequently used to treat wastewater with high suspended particle concentrations or to dewater sludge.

10.5.2   BIOLOGICAL TREATMENT

The biological treatment is also known as secondary treatment. It used the capabilities of microorganisms to break down organic materials in wastewater. It is an important and necessary part of any facility that manages wastewater from either a municipality or an industry with soluble organic contaminants, or from a mix of the two types of wastewater sources. There are two ways to carry out biological treatment such as aerobic and anaerobic biological treatments.

10.5.2.1　Aerobic Biological Treatment

In wastewater treatment, aerobic or facultative anaerobic bacteria are used to oxidize and decompose organic materials. Microorganisms' enzyme processes oxidize and break down organic materials, generating energy in the process. Microorganisms reproduce by using some of the energy and organic materials. Microorganisms that have developed in excess must be isolated and disposed of as extra sludge. To breakdown the organic materials by delivering air (oxygen) to the mixed liquor, wastewater is mixed with microbial flocs (activated sludge) in the aeration tank in an active sludge system. The sedimentation tank separates the mixed liquor, and the precipitated activated sludge is then sent back to the aeration tank. The water from the supernatant is released as treated water. The maintenance of an environment that is conducive to the action of microorganisms is crucial in the treatment of activated sludge. The effectiveness of removing organic materials is lowered or an inadequate separation of sludge from water (bulking) may occur when the environment changes and microbial activity decreases (Cheremisinoff, 2019; Mittal, 2011).

10.5.2.2　Activated Sludge

When the water is kept for 48 hours in the aeration tank, the impurities settle to the bottom of the tank and the top of the tank is filled with pure water. The activated sludge contains bacteria, exo-polymers of bacteria, and protozoans. These protozoans do not take place directly in the treatment but keep control of the bacteria population and are required in the process as they decide the efficiency of the treatment. The following are some aspects that contributed to the activated sludge's effectiveness:

- Formation of floc is readily aggregated and settles down (factors include oxygen, pH, etc.; 6–8 pH induces filamentous enzymes to proliferate and break down floc).
- Appropriate aeration is required.
- The optimum operating temperature should be 15–30°C.
- The ideal oxygen concentration is 1–2 mg/L.
- Protozoans can be used to determine the efficiency of this activated sludge.
- Succession of protozoans
  - (a) Amoeboid protozoans (they fed on the organic matter when there is no oxygen then the population will decrease as the bacteria start multiplying and increasing the population).
  - (b) Flagellated protozoans (they fed on the bacteria but when they lose in the competition between bacteria and flagellated protozoans).
  - (c) Frees swimming ciliates.
  - (d) Crawling ciliates.
  - (e) Stalked ciliates.

10.5.2.3　Anaerobic Biological Treatment

Anaerobic microorganisms are used in this treatment process, which also goes by the names anaerobic digestion or methane fermentation, to break down organic materials. Sludge or wastewater are added to a closed tank that is maintained in anaerobic

TABLE 10.2
Major Differences Between Aerobic and Anaerobic Water Treatment

Variables	Aerobic	Anaerobic
Principle	Biological activity that utilizes oxygen to degrade organic pollutants and other impurities, such as nitrogen and phosphorous	Biological activity that utilizes microorganisms and does not need oxygen to break down pollutants present in wastewater
Reaction product	Carbon dioxide, water and excess biomass	Carbon dioxide, methane and excess biomass
Reaction kinetic	Fast	Slow
Energy required	Relatively high	Relatively low
Efficiency	Relatively high (98%)	Relatively low (70–95%)
Applications	Municipal sewage, refinery wastewater treatment	Food and beverage wastewater rich in starch/sugar/ alcohol

conditions and occasionally warmed to help it digest. The tank's retention time ranges from a few days to tens of days. In general, anaerobic treatment is effective for handling wastes with high levels of organic materials.

Polysaccharides, lipids & protein

↓ Hydrolysis

Sugar, fatty acid & amino acid

↓ Fermentation

Acetate, CO_2, H_2

↓ Methanogenesis

$CH_4 + CO_2$ and $CH_4 + H_2O$

↓

Used as fuel to power water treatment plant

10.6 ADVANTAGES

Biological treatment methods provide a few advantages over other forms of treatment, including

- Treatment technique is established and well-known.
- Improved elimination of organic content effectiveness.
- Cost-effective, safe, and
- Environmentally friendly.

REFERENCES

Ball, P. (2017). Water is an activematrix of life for cell and molecular biology. *Proceedings of the National Academy of Sciences of the United States of America, 114*(51), 13327–13335. https://doi.org/10.1073/pnas.1703781114

Bezuneh, T. 2016. The role of microorganisms in distillery wastewater treatment: a review. *Journal of Bioremediation and Biodegradation, 7*, 375.

Braker, G., & Conrad, R. 2011. Diversity, structure, and size of N_2O-producing microbial communities in soils—what matters for their functioning? *Advances in applied microbiology, 75*, 33–70.

Cheremisinoff, P. N. 2019. *Handbook of water and wastewater treatment technology.* Routledge.

Craggs, R. J., Lundquist, T. J., & Benemann, J.R. 2013. Wastewater treatment and algal biofuel production. *Algae for biofuels and energy*, 153–163.

Delpla, I., Jung, A. V., Baures, E., Clement, M., & Thomas, O. (2009). Impacts of climate change on surface water quality in relation to drinking water production. *Environment International, 35*(8), 1225–1233. https://doi.org/10.1016/J.ENVINT.2009.07.001

Gay, M., Cerf, O., & Davey, K. 1996. Significance of pre-incubation temperature and inoculum concentration on subsequent growth of Listeria monocytogenes at 14° C. *Journal of Applied Microbiology, 81*(4), 433–438.

Grady Jr, C. L., Daigger, G. T., Love, N. G., & Filipe, C. D. 2011. *Biological wastewater treatment.* CRC press.

Jachimowski, A. (2017). Factors affecting water quality in a water supply network. *Journal of Ecological Engineering, 18*(4), 110–117. https://doi.org/10.12911/22998993/74288

Jasmin, M., Syukri, F., Kamarudin, M., & Karim, M. 2020. Potential of bioremediation in treating aquaculture sludge. *Aquaculture, 519*, 734905.

Lintern, A., Webb, J. A., Ryu, D., Liu, S., Bende-Michl, U., Waters, D., Leahy, P., Wilson, P., & Western, A. W. (2018). Key factors influencing differences in stream water quality across space. *Wiley Interdisciplinary Reviews: Water, 5*(1), e1260. https://doi.org/10.1002/WAT2.1260

Mittal, A. 2011. Biological wastewater treatment. *Water Today, 1*, 32–44.

Nielsen, F. H. (2000). Evolutionary events culminating in specific minerals becoming essential for life. *European Journal of Nutrition, 39*(2), 62–66. https://doi.org/10.1007/s003940050003

Ochoa-Herrera, V., Banihani, Q., Leon, G., Khatri, C., Field, J.A., & Sierra-Alvarez, R. 2009. Toxicity of fluoride to microorganisms in biological wastewater treatment systems. *Water research, 43*(13), 3177–3186.

Rani, N., Sangwan, P., Joshi, M., Sagar, A., & Bala, K. 2019. Microbes: a key player in industrial wastewater treatment. In: *Microbial wastewater treatment*, Elsevier, pp. 83–102.

Samocha, T. M. 2019. *Sustainable biofloc systems for marine shrimp.* Academic press.

Shahid, M. J., AL-surhanee, A. A., Kouadri, F., Ali, S., Nawaz, N., Afzal, M., Rizwan, M., Ali, B., & Soliman, M. H. 2020. Role of microorganisms in the remediation of wastewater in floating treatment wetlands: a review. *Sustainability*, **12**(14), 5559.

Show, K.-Y., & Lee, D.-J. 2017. Anaerobic treatment versus aerobic treatment. *Current developments in biotechnology and bioengineering*, 205–230.

Tao, M., He, F., Xu, D., Li, M., & Wu, Z. 2010. How artificial aeration improved sewage treatment of an integrated vertical-flow constructed wetland. *Polish Journal of Environmental Studies*, **19**(1).

Westall, F., & Brack, A. (2018). The importance of water for life. *Space Science Reviews*, *214*(2), 1–23. https://doi.org/10.1007/s11214-018-0476-7

Whitehead, P. G., Wilby, R. L., Battarbee, R. W., Kernan, M., & Wade, A. J. (2009). A review of the potential impacts of climate change on surface water quality. *Hydrological Sciences Journal*, *54*(1), 101–123. https://doi.org/10.1623/hysj.54.1.101

Index

A

Accurate/realistic system models, 32
Adaptive neuro-fuzzy inference system (ANFIS) models, 86, 91
Ag-BiVO$_4$-MnO$_x$, electron-hole pair formation, 108
Alkaline (CaCO$_3$), 96
Aluminium chloride, wastewater treatment, 81
Aluminium chlorohydrate, wastewater treatment, 81
Aluminium sulphate, wastewater treatment, 81
Ammonia, NH$_3$ oxidation, 185
Ammonia-oxidizing bacteria (AOB), 186
Anaerobic microorganisms, 185, 189
Anaerobic water treatment, 190
Anatase, electron-hole restructurization, 133
Anionic polyelectrolytes, wastewater treatment, 81
Ansys FLUENT$_{TM}$, 53, 123, 151
Aragonite, crystallization of, 96
AutoCAD, 157

B

Benzene 1,3,5–tricarboxylic acid chloride (*see* Trimethyl chloride (TMC))
Biofilm, 108, 185
Biological floc, 185
Blood glucose meters, 67
Boltzmann constant, 174
Boltzmann's equation, 152, 153
Boron nitride nanotubes (BNNTs), 25
Boundary conditions (BCs)
  in CFD simulations, 70
  drift diffusion, at different plasma model interfaces, 159
  Fick's law, 163
  in hydrodynamics, 69
Brinkman equation, 2D coupling of, 51
Brownian motion, 172

C

Cake formation, 170, 172
Calibration-library-interpolation (CLI) scale, 3
Carbon nanotube (CNT) membranes, 19
  layers of, 19
  water propagation, 19
Catalyst's ferromagnetic configuration, 99
Catchment
  distribution, 184

matter sources, 184
  mobilization, 184
Cefixime, in wastewater contamination, 145
Cellulose acetate, 14, 45
Charge conservation, 155
Charge density equation, 155
Charge neutralization, 79–80
Chemical contaminants, 185
Chemical oxygen demand (COD), 14, 105, 108
Chemical pollutants, 67
Chen, V, 27
Climate change, by humans, 183
Coagulant treatment (*see also* Coagulation)
  ANNs/ANFIS, coagulant dosage prediction model using, 90–91
  Bearspaw WTP SVM model, with kernel functions, 91–92
  case studies
    Ban Song WTP, 88
    Bearspaw Water Treatment Plant (WTP), 89
    Ceara water treatment plant, 89–90
    data neutralization/training, 89
    variable selection/criteria, 90
    water quality, variation of, 88–89
  charge-neutralizing component, 79
  dosage, models of, 80, 85–88
    adaptive neuro-fuzzy inference system (ANFIS) models, 86
    DTLFN model, 86
    Elman recurrent network (ERN), 86
    FTLNF model, 86
    jar testing, 83–84
    microscale dewatering tests, 84
    model performance, evaluation criteria of, 87–88
    nonlinear autoregressive with exogenous input (NARX) model, 87
    streaming current detectors (SCD), 84, 85
  hypothesis of, 82
  inorganic coagulants, 81
  methods of mixing, 82–83
  organic coagulants, 81
  PAC/AS dosages prediction, 92
  pH levels, 80
  water treatment plant (WTP), 80
    flowsheet of, 81
Coagulation
  basins, 83
  flocculation, 80
  hypothesis, 82

193

For Product Safety Concerns and Information please contact our EU
representative GPSR@taylorandfrancis.com
Taylor & Francis Verlag GmbH, Kaufingerstraße 24, 80331 München, Germany

www.ingramcontent.com/pod-product-compliance
Lightning Source LLC
Chambersburg PA
CBHW070713220326
41598CB00024BA/3130